Containment System Design

Chemical Storage, Mixing and Recycling

Fredric R. Haskett

Copyright © 1995

Previous edition copyright © 1991 by Fredric R. Haskett

Printed in the United States of America

10 9 8 7 6 5 4 3 2 1

ISBN 0-929870-33-6

Library of Congress Catalog Card Number 95-79496

Published by ADVANSTAR Communications, Inc.

ADVANSTAR Communications is a U.S. business information company that publishes magazines and journals, produces expositions and conferences, and provides a wide range of marketing services.

For additional information on any magazines or a complete catalog of ADVANSTAR Communications books, please write to ADVANSTAR Communications, Inc.; 7500 Old Oak Boulevard; Cleveland, OH 44130.

Cover design, interior design, editing and indexing: Lachina Publishing Services

Product Manager: Karen Eagle

Additional editing provided by Jerry Roche

Foreword

This guidebook contains information that allows both large and small applicators to design and construct an affordable, viable, and safe facility to store, handle, mix, and recycle the pesticides and fertilizers that are so necessary in our work.

The primary goal of this book is to explain the problems confronting our industry in regard to these issues and to offer a simple, cost effective, and safe way of dealing with them.

It is my hope that this book can give you the knowledge I've learned the hard way and that you can avoid problems detrimental to our businesses, institutions, and the industry.

Special thanks to Jim Brooks at The Lawn Institute, who encouraged me to speak and write about this issue, and to Henrietta and Beth, without whose help and encouragement it would not have been possible.

Fred Haskett

Contents

Summary Discussion   34

APPENDIX I

Equipment Standards   37

Site Standards   44

Emergency Response Procedures   60

APPENDIX II

Block Diagrams   67

A Special Introduction from the Editor-in-Chief of *Landscape Management* Magazine

If you are involved in mixing, spreading, spraying, or otherwise handling synthetic fertilizers and pesticides, this book is for you.

Landscape Management magazine first crossed paths with Fred Haskett years ago, when he was owner of a medium-sized lawn care business in Dover, Ohio. Though most lawn care operators, landscapers, and golf course superintendents are environmentally conscious, Fred's thinking was always a step ahead of his colleagues. Using that special foresight, he looked beyond the environmental reasons for a proper containment facility, to employee safety and lower insurance rates.

Even if Ohio environmental law hadn't been on the threshold of requiring approved pesticide facilities—which it was—Fred probably would have gone ahead with his plans. He's that conscientious.

Throughout his lengthy and costly ordeal, Fred kept excellent notes. Not long after he had finished designing and building his containment and storage facility, we let him use the pages of our magazine to introduce his story. We wanted to allow our 47,000 readers the opportunity to bypass some of the problems he had to overcome.

In the four years since we published that article, Fred has become a self-made "expert" and consultant on pesticide containment and storage systems. He recently formalized his original notes into this book. We at *Landscape Management* magazine and the Advanstar Communications Marketing Services Division are proud to share his revelations with you. And we congratulate our friend and associate Fred Haskett on his commitment and dedication to the green industry.

Jerry Roche

Introduction

As a result of experiences gained over a twenty-one-year career, Fred Haskett has written a guidebook entitled *Containment System Design.* Formerly the owner/operator of Greenworld Landscape Management in Dover, Ohio, he has since 1991 served as Director of Operations for Ehrlich Lawn and Tree Care. Headquartered in Reading, Pennsylvania, Ehrlich's is a diverse service company providing lawn and tree care, with thirty-six district offices in the six-state mid-Atlantic region.

In 1985 and 1986 Haskett designed, developed, and installed a comprehensive integrated system for safe storage, efficient mixing and handling, spill containment, and residue recycling that has proven to be both practical and affordable. In four years of operation, the system has proven itself in several ways, such as increased employee safety; reduced insurance rate growth; eliminated or significantly reduced cost and liability of both storage and disposal of waste residues; and ease and affordability of maintenance.

This system was examined and approved for use by the Ohio Environmental Protection Agency, the Ohio Department of Agriculture, and Dover, Ohio's water, wastewater, and health departments.

The system is flexible and can easily be used in new construction or retrofitted into an existing structure. It can be readily customized to suit virtually any operation's varied needs and economic priorities, no matter how large or small a particular facility is. The system's basic simplicity and adaptability suit it to any type of pesticide user, including lawn or landscape firms, golf courses, sod farms, nurseries, pesticide control firms, and others.

The purpose of this guidebook is to show the reader in a general way some of the available options for designing and developing a system. It is, therefore, the reader's responsibility to ensure that any system designed and constructed is in compliance with federal, state, and local regulations specific to the location.

It is hoped that the information provided herein will be of value to the reader, because so little data has been heretofore formalized on this subject. The provisions of this guidebook should not be construed to mean that no outside assistance with design and construction is necessary, particularly concerning the variation in state and local regulation and the changing federal regulations both in effect and pending as of this writing.

Overview

If a company handles and stores pesticide concentrates and fertilizers in any quantities, and does not have an integrated system for the safe storage, handling, and residue recycling of these products, it could be risking its business every day.

Industries that use pesticides and fertilizers are being watched by a host of different groups such as federal and state environmental protection agencies, federal and state departments of agriculture, local governments, environmental groups, the media, the insurance industry, and the public at large. Currently, laws requiring containment systems are on the books in thirteen states: Illinois, Indiana, Iowa, Kansas, Michigan, Minnesota, Missouri, Nebraska, Oklahoma, Ohio, South Dakota, West Virginia, and Wisconsin. Florida and Idaho have established special guidelines for storage and containment. In addition, eight states have bills in process before their legislations: California, Colorado, Kentucky, Louisiana, North Dakota, Tennessee, Texas, and Washington. It is only a matter of time before more states join this group. These new laws, in addition to civil penalties and fines, contain the provision for criminal penalties and jail terms. Also, the current Resource Conservation and Recovery Act amendments have reduced the minimum waste generation and storage quantities drastically, thus mandating compliance for all but the very smallest pesticide end users.

The underlying questions are the following:

1. Can your facility stand up to this scrutiny?
2. Can your operation comply with these new regulations?
3. Do you know the legal and economic implications of compliance or noncompliance?

1

4. Are you aware of the cost difference between recycling residues and rinsates and having them disposed of properly?

If the answer is no to any or all of the above questions, the results could be catastrophic. You could face retroactive fines, lawsuits, criminal and civil penalties, jail terms, negative publicity, large rate increases or loss of insurance, enormous cleanup or disposal costs, or the loss of your business. One estimate of the cost of residue and rinsate disposal for the average facility ranges from $8,000 to $15,000 annually. In addition, cleanup costs of spills as low as one gallon of concentrate can range from $50,000 to $100,000.

The industry has done a good job of countering the negative issues raised in the areas of consumer safety and application safety. Also, the risks versus the benefits of various containment services has been well documented.

However, the industry has not done a very good job in interior operations at the point source locations for serious potential incidents, including concentrates and waste materials. Current estimates are that 80 to 90 percent of pesticide end users are not prepared to implement systems or procedures for the new regulations; nor are they ready for or capable of dealing with the economic consequences of noncompliance.

In order to comply successfully with the new or proposed regulations and to provide effective protection from serious accidents, the following issues must be addressed:

- Safe storage of concentrated pesticides
- Safe mixing and handling systems and procedures
- An effective system for the recycling/reuse of excess or waste concentrates and dilute residues

The first issue to examine is how and where pesticide concentrates—both liquid and dry—are stored. This area should be compartmentalized into two distinct sections: a primary containment and a secondary containment area.

The primary containment area is used for both storage and mixing operations involving concentrated pesticides. This area must be diked, and the floor and dikes treated and/or coated with watertight, wear-resistant materials that are also resistant to chemical corrosion. Such treatment ensures that any spills generated within this area can be contained and recovered rather than released into the environment. In addition, this area should be further segregated by a partition to control unnecessary or unauthorized access. Equipment such as spill recovery tools, emergency shower/eye wash, and fire extinguishers is also an integral part of this section. Finally, a ventilator fan is mandatory to clear out fumes, dust, or gases present during storage or mixing operation.

The secondary containment area is used for storage and parking of spray rigs or trucks, for loading or fill operations with dilute pesticide mixes and fertilizers, and for washing and rinsing of pesticide residues from application equipment and vehicles. Dry fertilizers can also be stored in this area. In addition, the storage tanks for recyclable dilute pesticide residue and rinsates are located here.

The secondary containment area can be sealed from adjacent areas with partial dikes at doorways and with floor coatings and watertight, wear-resistant wall coatings that are resistant to chemical corrosion. Once again, ensure that any spills or discharges within this area can be contained, recovered, and not released into the environment. The equipment used here is identical to that used in primary containment.

Following these basic concepts can help in achieving safe and efficient storage, mixing, loading, and cleanup. An effective integration of the primary and secondary containment areas and their systems can reduce fill times, while at the same time reducing opportunities for mishandling, accidental spillage, unnecessary exposure to staff, and waste of expensive pesticides.

The recovery and recycling of pesticide containment wash water, rinsates, dilute residues, and waste concentrates are one of the most critical aspects of this operation. An effective and comprehensive recycling system can be one of the most important systems for protecting any business from becoming a storage site for hazardous waste. In addition, the astronomical costs for disposing of these materials in accordance with federal and state regulations can be greatly reduced or eliminated.

One can significantly reduce or even eliminate the generation of any hazardous chemical waste through the consistent daily or weekly use of (a) a three-way mixing program of concentrated pesticides, fertilizers, and adjuvants as required; (b) compatible residues and rinsates diluted with 90 percent fresh water; and (c) fresh water to make up the balance of fill. More importantly, one can avoid the extreme liabilities and costs that accompany such a situation.

Finally, a successful system offers protection by keeping hazardous materials away from outside water systems. The primary protection is the use of a back-flow prevention device at the main source. In addition, all water outlets, with the exception of restrooms, are to be equipped with anti-siphon devices for backup protection.

Additional protection for exterior groundwater areas and for sanitary sewer and storm sewer systems is achieved with an integrated combination of containment dikes, self-contained recovery sumps, and a system of coatings on the floor and walls.

There is absolutely no reason that every end user of pesticides, no matter how small, should not have a similar system. Sooner or later there will be no choice, and those companies who have dealt with the problem and solved it will be better off than those who buried their heads in the sand.

So why comply? What are the benefits? There are essentially three benefits: economic, environmental, and employee safety.

Economic benefits include (1) avoiding the high cost of chemical waste disposal once full enforcement is in place, which could cost a small company $8,000 to $10,000 annually; (2) reduction of the exposure to spills and accidents and the potentially astronomical cleanup costs; (3) savings and/or reduction in insurance premium rate growth; and (4) tax credit for the installation and maintenance of the anti-pollution system.

Environmental benefits include (1) the reduction and/or elimination of point source contamination to groundwater, sewer, and water systems; and (2) elimination of the storage, transportation, and disposal of chemical waste residues.

Employee safety benefits include (1) improved safety; (2) ease of operation; and (3) increased feeling of well-being from association with the system.

In the final analysis, there are two choices—compliance or evasion. The regulations are here or on the way, and those who act appropriately will have no trouble with them. The implications for those who evade will be enormous, with potential for fines, shutdowns, other criminal and civil penalties, negative publicity, lawsuits, and waste.

A system such as this should be used for several reasons: (1) it works—after four years of operation, we know it works; (2) it has been approved by the EPA and the Ohio Department of Agriculture; (3) it can be used with new construction or retrofitted into any existing building; (4) it is easy to install, operate, and maintain; (5) it is affordable to install and uses available materials; and (6) most importantly, one can avoid the stigma of the label "hazardous waste storage site" and the accompanying high cost of proper disposal.

Questions and Answers: Direction

Q. Why should the industry increase its emphasis on environmental protection and safe handling of pesticide products?

A. Because companies use agricultural chemicals that are effective and safe when handled properly. They do not want products to reach groundwater or to be used in an unsafe manner. It is in the best interests of the industry, as well as individual companies and the public in general, to exert every effort to ensure that pesticide products are handled carefully. Companies want employees to understand their intentions and to provide an example to customers and the public. This example shows the importance of safe practices and respect for the environment.

Q. Won't the high standards required for the containment and recycling systems cost employers business?

A. Those in the agricultural chemical industry should evaluate their business in terms of the future as well as the present. This evaluation must include the safe handling of products to protect the environment. Many locations have already invested large sums of money to protect employees and the environment. Those who have not already made these investments will need to do so in the future because of increased regulatory requirements and liability concerns. Those locations that are either unable or unwilling to make these commitments may not be able to do business in this industry in the future. This should be a conscious business decision. In addition, the annual cost of proper disposal could exceed the one-time installation cost.

Q. If a state has not yet finalized containment standards, why should a company make the financial site commitments?

A. Some manufacturers and distributors will not sell products to locations that do not meet minimum standards. One should thoroughly discuss the situation with them to take advantage of any programs and technical advice that can help meet minimum standards. Other resources within the industry can be contacted to determine the status of proposed regulations and standards.

Q. What if a company refuses to meet minimum site standards in order to make its goal? What should it do?

A. Fully investigate the situation. There are strong arguments to convince the company if it is committed to staying in the business long term. If the company is unable to see the benefits of responsible environmental actions, it should develop a plan to meet its goals in another business.

Q. What are a company's environmental responsibilities? How accountable is it?

A. Safety and environmental issues, including consistent application of containment and recycling site standards, are an important part of a company's operation. These are as important as sales activities. When a company certifies that a site is in compliance, it is stating that these standards have been met.

The application of site standards is sometimes complex, and it is difficult to fully evaluate a site. Mistakes are sometimes made; sometimes a location commits to doing something that does not happen. However, *intentional* misrepresentation places the company in a vulnerable position and could have serious consequences, such as civil and criminal penalties, fines, and jail terms.

Q. The large corporations, the manufacturers, and the industry associations make statements asserting their environmental commitment, but what *actions* support that commitment?

A. A number of actions can be cited: product development standards that screen for environmental concerns; establishing the Environmental Task Force (regional and national); an aggressive effort at the legislative level for reasonable regulations; and so on. The importance of site safety cannot be overemphasized.

One of the reasons for this guidebook is to help coordinate and make consistent a standard of safe and efficient operation. Another is to make it more affordable to comply with the new and forthcoming regulations.

Planning

Good planning will result in a facility that operates smoothly, permits flexibility and growth, and protects the environment.

The facility should be situated to receive, as well as dispatch, vehicles of all sizes associated with the type of operation.

When selecting or evaluating a site, examine the area's topography and its relation to streams, bodies of water, storm sewers, sanitary sewers, and fresh water systems. With this information one can begin to develop alternatives for spill containment.

When a facility is newly constructed or retrofitted, it is advisable to test the soil and water at the site to ensure that there is no existing contamination.

One must develop a contingency plan for handling potential accidents or emergencies.

All components of a pesticide storage/handling/recycling system should be carefully selected for chemical compatibility with the products that will be handled. Valves, lines, gaskets, and fittings must be strong and well made to avoid problems during operation.

The system should be designed so that it can be serviced easily. Major components, such as pumps and meters, should be installed with shut-off valves so that they can be removed while the system is full. Valves should be placed before and after each component and attached to the plumbing with flanges on unions. Components should also be mounted so that they can be drained if necessary.

Dikes, wash pads, floors, and walls should be treated and/or coated with watertight, wear-resistant materials that are also resistant to chemical corrosion.

The system and the facility should be designed to meet the needs of the specific operation, now and in the future.

Tanks, delivery systems, components, and facilities must comply with federal, state, and local regulations.

Storage tanks, transport tanks, meters, pumps, and lines must be free of contaminants before they are used to store or handle any chemical or chemical mix.

Personnel who handle pesticides should wear appropriate clothing and protective equipment as designated on the label or safety data sheet for the product or products in use. Personnel should be trained in safe operational and emergency procedures.

Storage, mixing, and loading areas should be equipped with an exhaust fan that is directed outside for ventilation of dust and fumes.

The following is a checklist to assist in the evaluation of an operation and its possible needs.

General Information

1. What is the type of business (for example, lawn care, golf course, pest control)?
2. Are pesticides used in the operations?
3. Are copies of current state and federal pesticide regulations maintained?
4. Are phone numbers of state and federal pesticide officials available?
5. Are copies of labels maintained for pesticides used at the facility?
6. Are copies of material safety data sheets (MSDSs) maintained for pesticides used at the facility?
7. Are labels and MSDSs centrally filed and easily accessible?
8. Are training manuals/materials kept on hand?

Applicator Certification

9. Does this facility employ one or more certified applicators?
10. Does this facility employ one or more noncertified applicators?
11. Does this facility have a reminder system to prevent accidental expiration of certification?
12. Do pesticide applicators receive training or attend recertification/continuing education courses?

Storage

13. Are pesticides stored in a separate room or building?

14. Is flood water or runoff water unlikely to enter the pesticide storage facility?

15. Does the storage facility have a drainage system with catch basin or aboveground holding tank?

16. Is the storage site secured by a climb-proof fence?

17. Are doors and gates at the site kept locked to prevent unauthorized entry?

18. Have warning signs been placed on interior and exterior exits, rooms, buildings, and fences at the site?

19. Have the local fire department, police, and other emergency response teams been notified of the types, location, and amounts of pesticides present at the facility?

20. Do local emergency response agencies have a floor plan of the storage facility showing pesticide locations?

21. Do local emergency response agencies have access to keys to the pesticide storage facility?

22. Do local emergency response agencies have the home phone numbers of the persons responsible for the pesticide storage facility?

23. Are combustible materials stored away from steam lines and other heat sources?

24. Are glass containers stored away from sunlight?

25. Is the pesticide storage area equipped with a fire alarm system?

26. Are pesticide-contaminated wastes stored securely to prevent unauthorized access?

27. Are pesticide-contaminated wastes stored separately from pesticides and other chemicals?

28. Are all pesticide containers stored with labels plainly visible?

29. When an original pesticide container is damaged, are contents transferred to a container that held exactly the same pesticide?

30. Are open bags of wettable powders, dusts, or water-dispersible granules stored in plastic bags or other sealed containers?

31. Are pesticides kept separate from each other according to class (e.g., herbicides, fungicides, nematicides)?

32. Are all containers stored off the ground in an orderly way?

33. Are breakable containers stored on lower shelves?

34. Are aisles provided to ensure effective access?

35. Is a complete inventory of pesticides recorded and kept current?

36. Are pesticide containers checked regularly for corrosion and leaks?

37. Are all electrical devices in the storage area shielded against sparks (i.e., explosion proof)?

38. Does the storage area ventilation exhaust to the exterior of the facility?

39. Can the storage facility be heated or cooled to avoid temperature extremes?

40. Are the telephone numbers of local emergency response teams posted at the storage facility?

Mixing/Loading

41. Do applicators review pesticide labels before handling or using pesticides?

42. Are mixing and loading operations always performed at the same location?

43. Is the mixing/loading area paved with impervious material and treated and/or coated with watertight, wear-resistant materials that are also resistant to chemical corrosion?

44. Is the mixing/loading area sloped away from water sources?

45. Is runoff from the mixing/loading area contained so as not to enter drainage systems or contaminate soils?

46. Is there a supply of clean water, soap, and paper towels available at the mixing/loading area?

47. Are adsorptive clay, kitty litter, absorbent pads, or soil available in case of spills or leaks?

48. Is there a check valve or anti-siphon device installed on the fill hose and other water outlets?

49. Do applicators wear protective clothing and equipment during mixing and loading?

50. Are written instructions provided to applicators at all times by appropriately certified individuals?

51. Is there a back-flow device installed on the incoming fresh water source?

Recycling

52. Are dilute residues, rinsates, and wash waters disposed of in accordance with state and federal regulations?

53. If not disposed of according to regulations, are they diluted with a 9-to-1 ratio of fresh water and used as makeup water and reapplied in approved usage?

54. Are all dilute residues and rinsates stored separately according to end-use compatibility?

The Containment Facility

The ideal containment facility has a practical, economical design that meets with regulatory approval, enhances operation efficiency, reduces waste, and improves worker and community safety. Such a system helps ensure long-term compliance and encourages investment in improved facilities.

The functions that are coordinated within this containment facility include the following:

1. Security, storage, handling, mixing, and loading of pesticides
2. Spraying rig cleanout and rinsing
3. External vehicle/equipment washing
4. Security and separation of concentrates and rinsates
5. Handling, storage, reuse, or disposal of rinsates and wash water, and containment of pesticide and fertilizer leaks and spills

Design Objectives

1. To develop a simple system suitable for loading, washing, and rinsing pesticide/fertilizer application equipment.
2. To provide areas for containing pesticide or fertilizer leaks and spills.
3. To incorporate security, storage, handling and mixing, loading, recovery, reuse, and disposal functions.
4. To use watertight, wear-resistant materials that are resistant to chemical corrosion, ground and weather stress, and mechanical damage.

5. To include separate self-contained sumps for concentrates and rinsates that are watertight, wear resistant, and resistant to chemical corrosion and ground and weather stress.

6. To provide for worker safety and ease of operation.

7. To provide flexibility and adaptability for future growth.

Facility Use

The facility should have two major sections: (a) a primary containment where concentrated pesticides are stored, handled, and mixed, possibly including a subdivision for liquid fertilizers; and (b) a secondary containment where sprayer washing, loading, and rinsing takes place, possibly including an area where dry fertilizer products can be stored away from water sources and use areas.

Primary Containment

The primary containment area is subdivided into three functional areas: a chemical storage area, a handling and mixing area, and a rinsate/liquid fertilizer storage area. The entire area is surrounded by a diked partition. In addition, the three subdivisions are separated by diking. Security is maintained through use of a six-foot-high chain-link fence with locked gates mounted outside the dike. Primary containment sizes are dictated by the equipment, containers, material, and traffic flow for each operation.

The liquid fertilizer/rinsate storage area should provide space for several holding tanks. The operation's needs should dictate the sizes. 300, 500, and 1,000 gallons are the norm for rinsate, while 1,000 to 5,000 gallons are typical for liquid fertilizers. The storage tanks should be mounted 3 to 5 inches above the floor to allow airflow and washing and to more readily detect leakage.

Within the handling and mixing area, the pumps should be placed on elevated platforms above liquid levels. An emergency eye-wash shower should be located here. An exhaust fan should be located in this area to clear away fumes and dusts from within the primary containment.

The chemical storage area is separated by another chain-link fence with a locked gate to limit access to authorized personnel only. All shelving and work tables within this area and the handling and mixing area should be polyethylene, fiberglass, or stainless steel. Wood is not acceptable.

Each of the primary containment subdivisions should have a separate self-contained sump to facilitate cleanup, spill recovery, and rinsate resi-

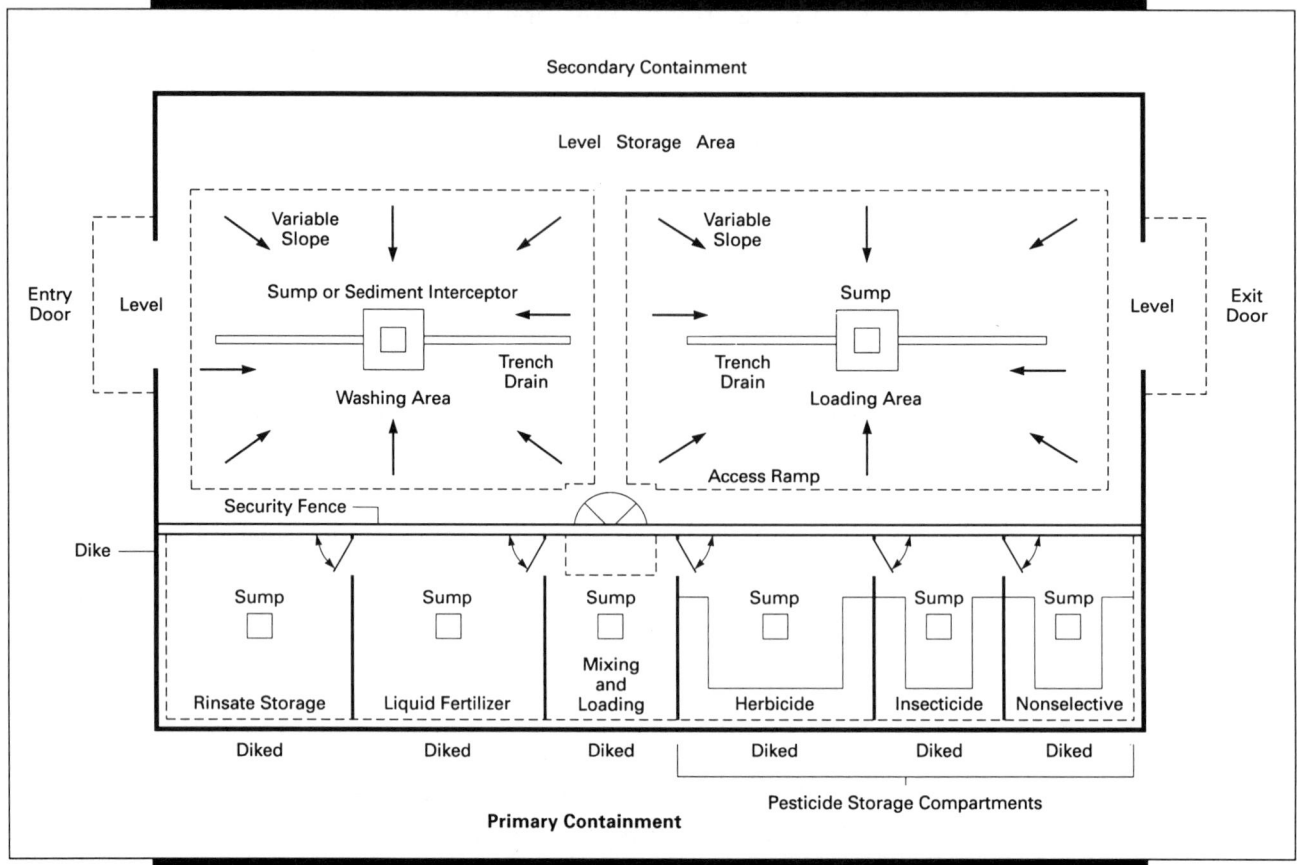

Plan View of Full-Service Facility Detailing Primary and Secondary Containment, Concentrate Storage, Mixing and Loading Area, and Rinsate Storage

due. All water sources entering this area should have hose connection breakers or atmospheric vacuum breakers as a standard item. In addition, a sink located within the handling and mixing area for cleanup of gloves, aprons, boots, and measuring equipment should drain into the sump for recycling.

Secondary Containment

The secondary containment area is subdivided into two basic areas: the exterior (road grime, etc.) vehicle wash area and the vehicle loading/contaminated equipment rinse area.

The exterior wash area is the first section entered. Positive control of exterior wash water (not considered a hazardous waste by EPA) can be maintained by a $2\frac{1}{2}$–$3\frac{1}{2}$ percent slope to a trench drain. The trench drain flows into a soil and sediment interceptor and then into a sanitary sewer or approved septic system.

The next section, or loading/rinsing area, is also protected by a $2\frac{1}{2}$–$3\frac{1}{2}$ percent slope along the center trench drain. Both of these sections slope to center trench drains. The outer pad edges are level, with the cross slope increasing from level at the edges to the maximum slope at the drains. The loading zone drains are channeled into a separate self-contained sump. The contaminated water is then transferred into the rinsate storage tank for use as makeup water.

The loading lines for mix chemicals, fresh water, makeup water, and so forth are overhead and always cross over containment areas. All fresh water lines are equipped with atmospheric breakers to prevent back flow. Loading lines for different target areas (for example, turf, ornamental, non-selective), both finish mix and makeup water, are located in distinct places and established to avoid cross contamination.

The divider dike separating the two sections has no openings; thus no spills or leakage can be inadvertently discharged into the sanitary sewer. It is designed to be driven over (like a speed bump).

Functional and Structural Design

Concrete floors in frequent contact with corrosive substances should be designed with a minimum 2 percent slope to facilitate washing. This design uses $2\frac{1}{2}$–$3\frac{1}{2}$ percent slopes in both containment areas. All separate sections slope to a center trench drain and a sump. The slopes increase from level at the edges to the maximum at the drains. Sumps should have tapered bottoms to facilitate pumping out liquids.

A watertight concrete design must be used to avoid leakage from dikes, sumps, drains, and the containment floor areas. These are to be considered as liquid holding and transfer vessels. Concrete mixtures for watertight construction should include the following:

1. A stiff dry mix for maximum strength, chemical and freeze resistance, and watertightness.
2. Type II to II-A cement with air entrapment at 4,000 to 4,500 psi compressive strength.
3. Concrete plasticity admixture for easier workability and improved watertightness and strength.
4. Vibration at 5,000–15,000 RPM.
5. Immersion or moist cure for 14 days minimum.

6. 1"–1.5" clean impervious aggregate.

7. Continuous pour in one day—no cold joints.

8. Install control joints at 12–16-foot intervals in both directions.

After completion, the following sealer materials should be applied: a two-stage penetrating sealer followed by a finish coating to drains, sumps, floors, and dike surfaces; and a silicone or Hypalon caulking to control seams and stress cracks. These sealer materials remain flexible after curing, aging, and weather stress and are resistant to chemical corrosion.

These guidelines and recommendations can also be applied to retrofitting existing facilities. By adding curbs or dikes to preexisting concrete floors, one can provide for storage, mixing and handling, rinsing, or washing. However, shallow slopes or level floors may not allow for adequate drainage to wash or recover spills. The main problem with concrete deterioration is the action of spilled fertilizer or pesticides on the surface. Spills or leaks can cause severe surface deterioration in just a few weeks if surfaces are not immediately and thoroughly cleaned. It is imperative, therefore, that these concrete surfaces be properly sealed, coated, and caulked. In addition, all dikes or curbs poured onto existing floor areas should have a concrete welding or bonding material applied between the old and new concrete. Seams should also be caulked and the dike sealed and coated.

Another critical factor in concrete pad site selection and preparation is to make sure that all surface water drains away from the pad. Soil moisture must not vary significantly year round. Excessive moisture combined with sub-freezing weather can cause severe heaving, adding excessive stress to concrete pads.

Capacity Design

The major problem in designing containment systems is determining the best combination of containment area and dike height. The containment holding volume should be designed to provide a minimum of 125 percent of the largest tank within the specific containment area or section. The ideal capacity is 125 percent of the total storage volume within the individual sections. With the right format and planning, this is possible at an affordable level.

Rinsate tank management systems are operated on average with three to six tanks ranging from 100–500 gallons each. Liquid fertilizer tanks typically range from 3,000–10,000 gallons each. Pesticide concentrates are stored in 1- to 5-gallon cans, 30- to 55-gallon drums, or 110-, 220-, or 330-gallon mini-bulk containers.

The area displaced by *all tanks,* including the area of the largest tank, plus any equipment in the containment area, must be added to the net fluid volume that can be released by the largest tank when the liquid stabilizes. Containment widths and sidewalls or dike heights need to be sized to accommodate this volume.

Special Notes

1. Total containment volume must be reduced by the volumes displaced by tanks and equipment that take up containment space.
2. Pesticide rinsate should not be stored in the same containment section with fertilizers.
3. Fertilizers must not, *by law,* be stored in the same containment section with full-strength pesticides!

Functional Layout

Storage tanks should not be installed too close to the containment section end walls or back side walls because of the possibility of a leak from the tank side wall. A lower side wall with a larger containment area provides more safety against leaks that might squirt over the side wall. Vertical tank seams should be oriented toward the center of the containment area.

Functional Layout Questions

1. Is there adequate space for present *and future* tanks, plus mixing and transfer equipment in the proposed containment area?
2. Will the area be adequate for potential growth, and can small existing tanks be replaced with larger tanks in the future?
3. From a safety standpoint, can workers move between tanks and walls with hoses and without undue risks or hazards?
4. Can all outside surfaces of tanks be visually inspected for corrosion, damage, and potential leaks?
5. Are tanks securely anchored or braced to prevent flotation, tipping over, and damage to other tanks?
6. Are tanks plumbed individually or with flexible plumbing connections to avoid plumbing damage and multiple leaks from tank flotation (no rigid pipe manifolds connecting tanks)?

7. Are gate and aisle widths adequate for safe and efficient movement of the largest portable equipment and containers?

8. Will access to vehicles, spray rigs, and equipment be safe and efficient?

9. Do workers have clear and easy access to safety equipment, such as emergency eye washes and showers, spill cleanup equipment, and fire extinguishers?

These questions should all be answered "yes" when the layout design is checked for tank fit, equipment fit, material movement, vehicle movement, workers' safety, and spill prevention.

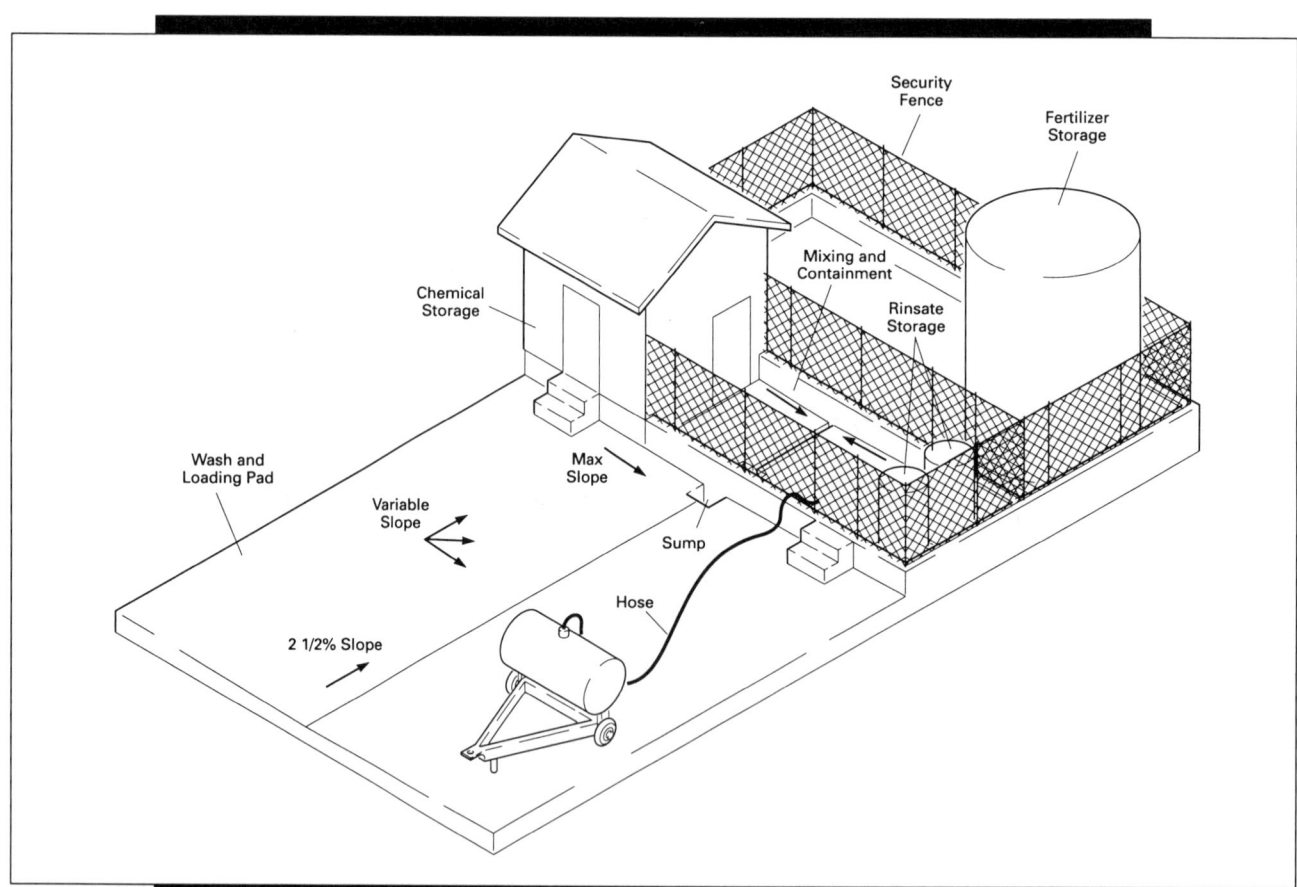

Schematic of Fertilizer, Pesticide Storage, Mixing/Loading Pad

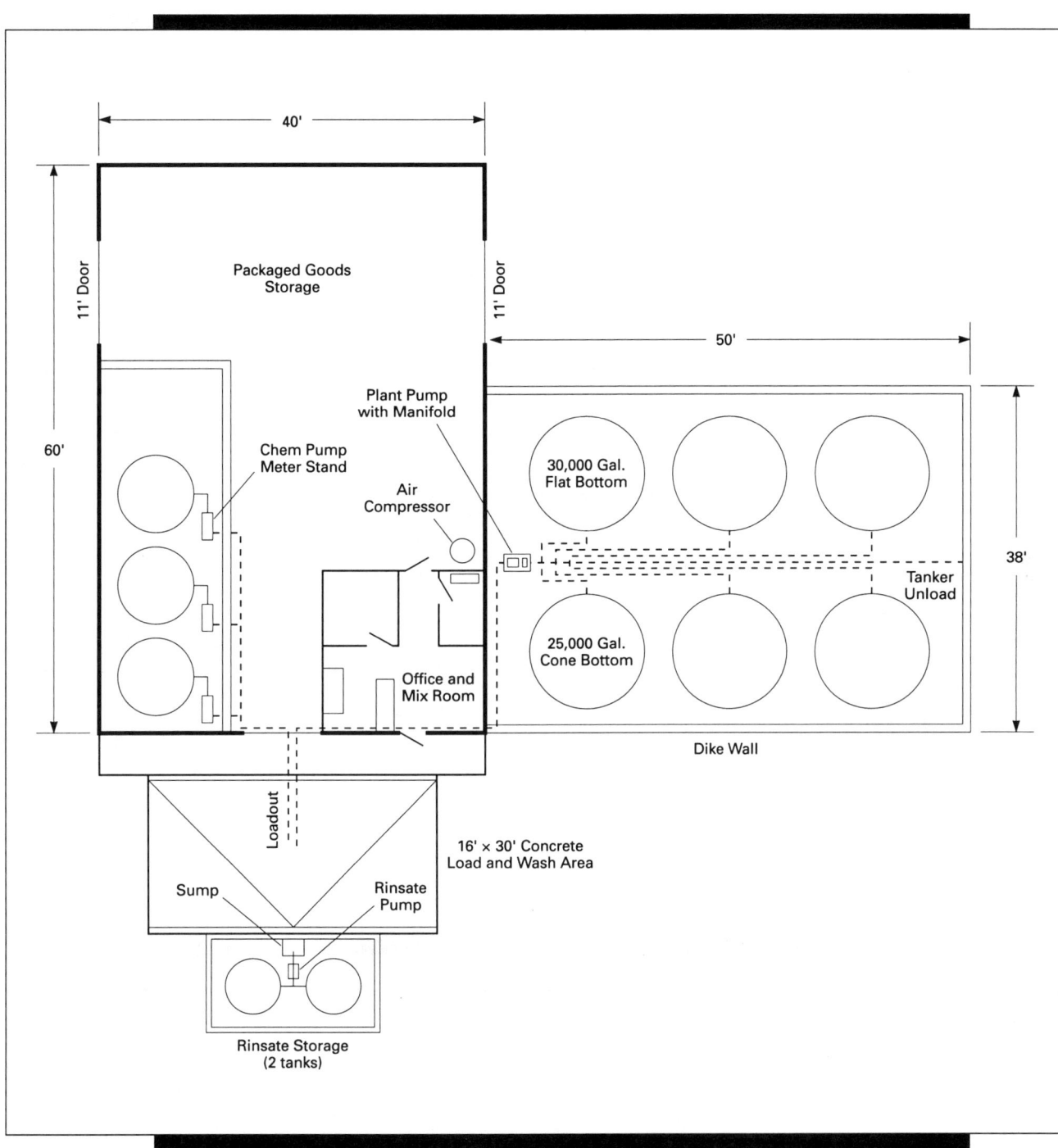

40'

11' Door

11' Door

Packaged Goods
Storage

50'

60'

Plant Pump
with Manifold

Chem Pump
Meter Stand

Air
Compressor

30,000 Gal.
Flat Bottom

38'

Tanker
Unload

Office and
Mix Room

25,000 Gal.
Cone Bottom

Dike Wall

Loadout

16' × 30' Concrete
Load and Wash Area

Sump

Rinsate
Pump

Rinsate Storage
(2 tanks)

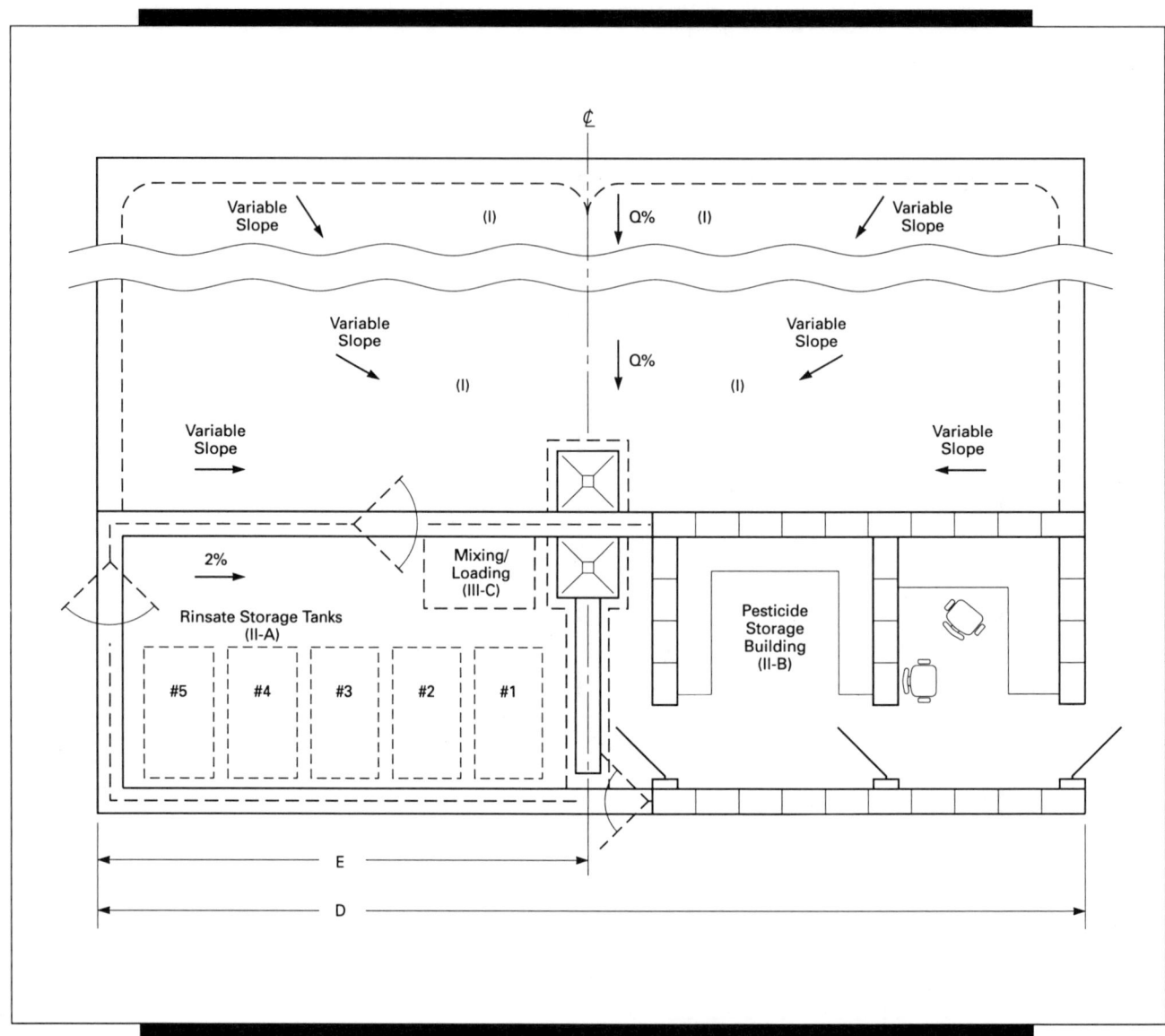

Plan of Containment Area with Pesticide Storage Building, Rinsate Storage, Mix/Load, and Security Fence

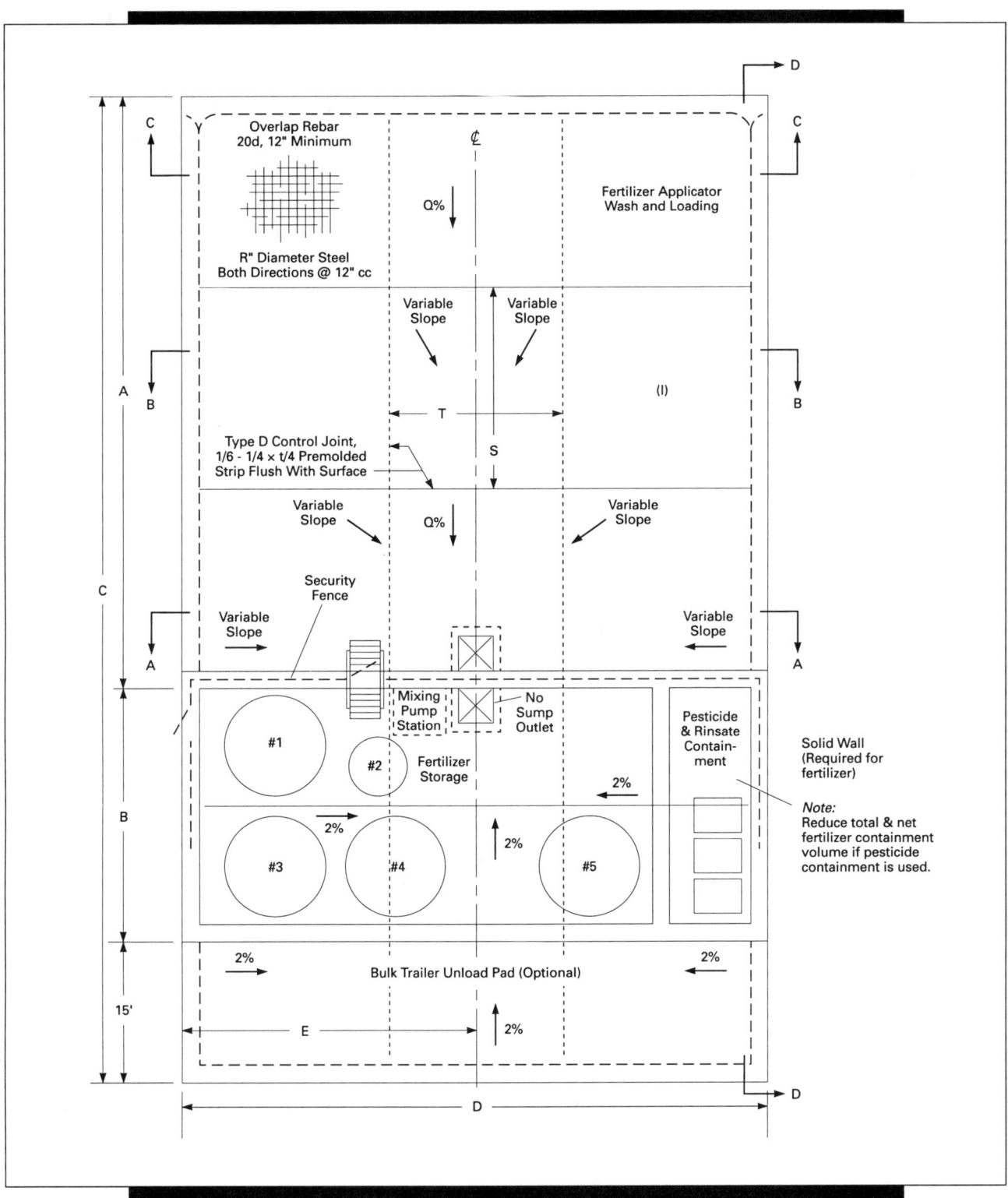

Plan View of Liquid Fertilizer and Pesticide Storage Facility with Optional Bulk Trailer Unload Pad

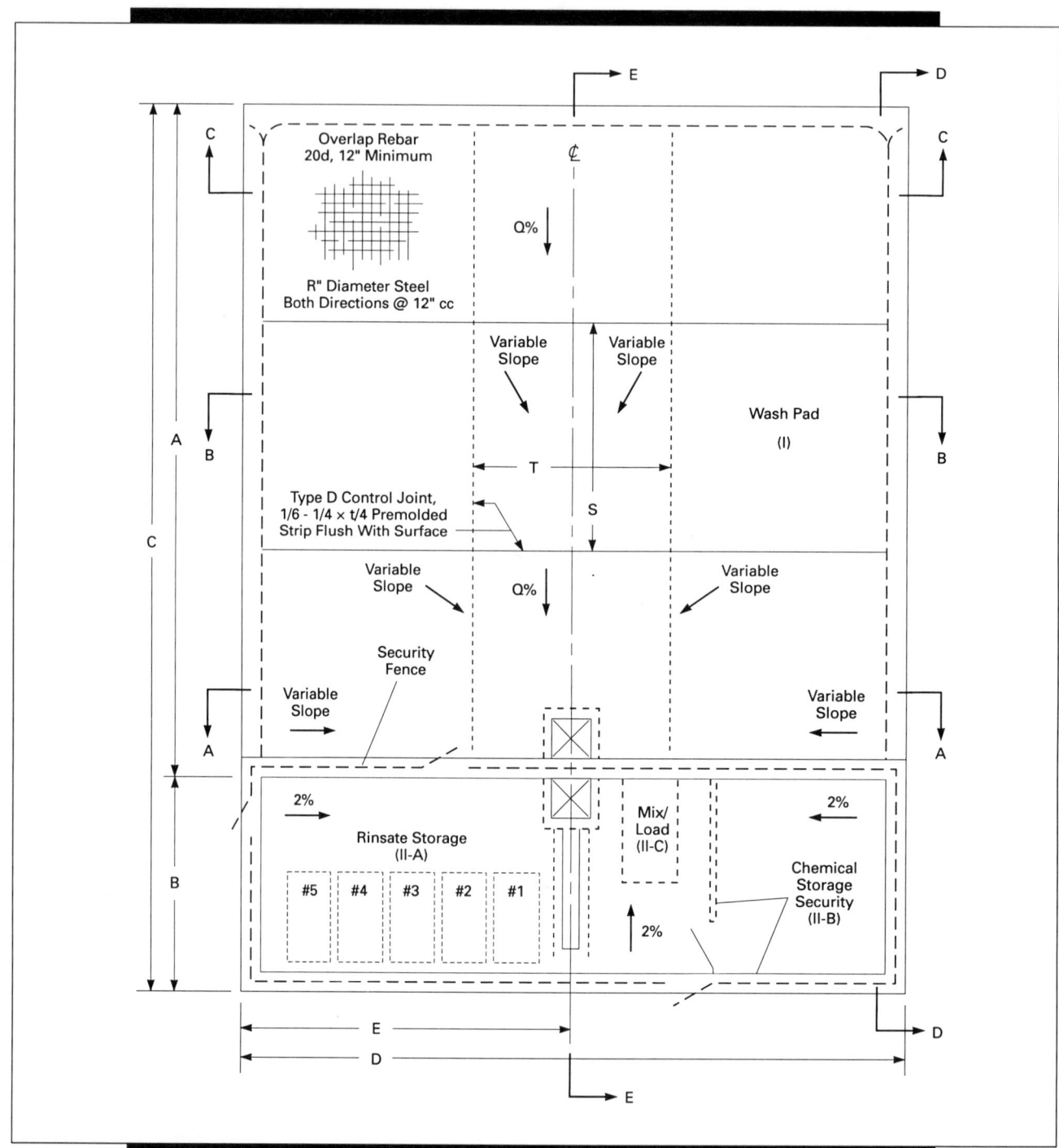

Plan View of Modular Concrete Vehicle Wash/Load Pad with Pesticide Containment

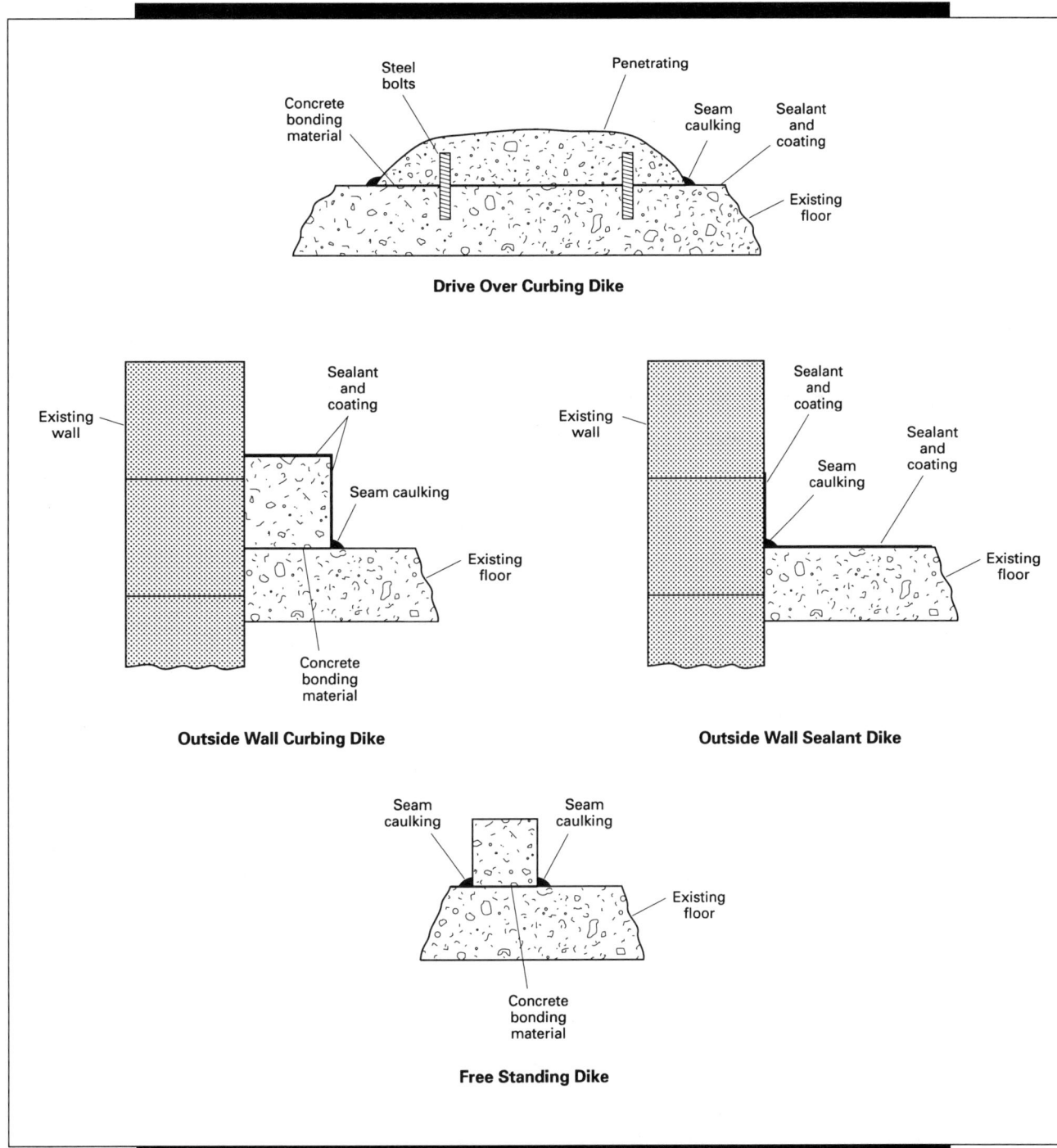

Drive Over Curbing Dike

Outside Wall Curbing Dike

Outside Wall Sealant Dike

Free Standing Dike

Cross-Section of Dike Configurations for Use with a Retro-fit Installation in an Existing Facility

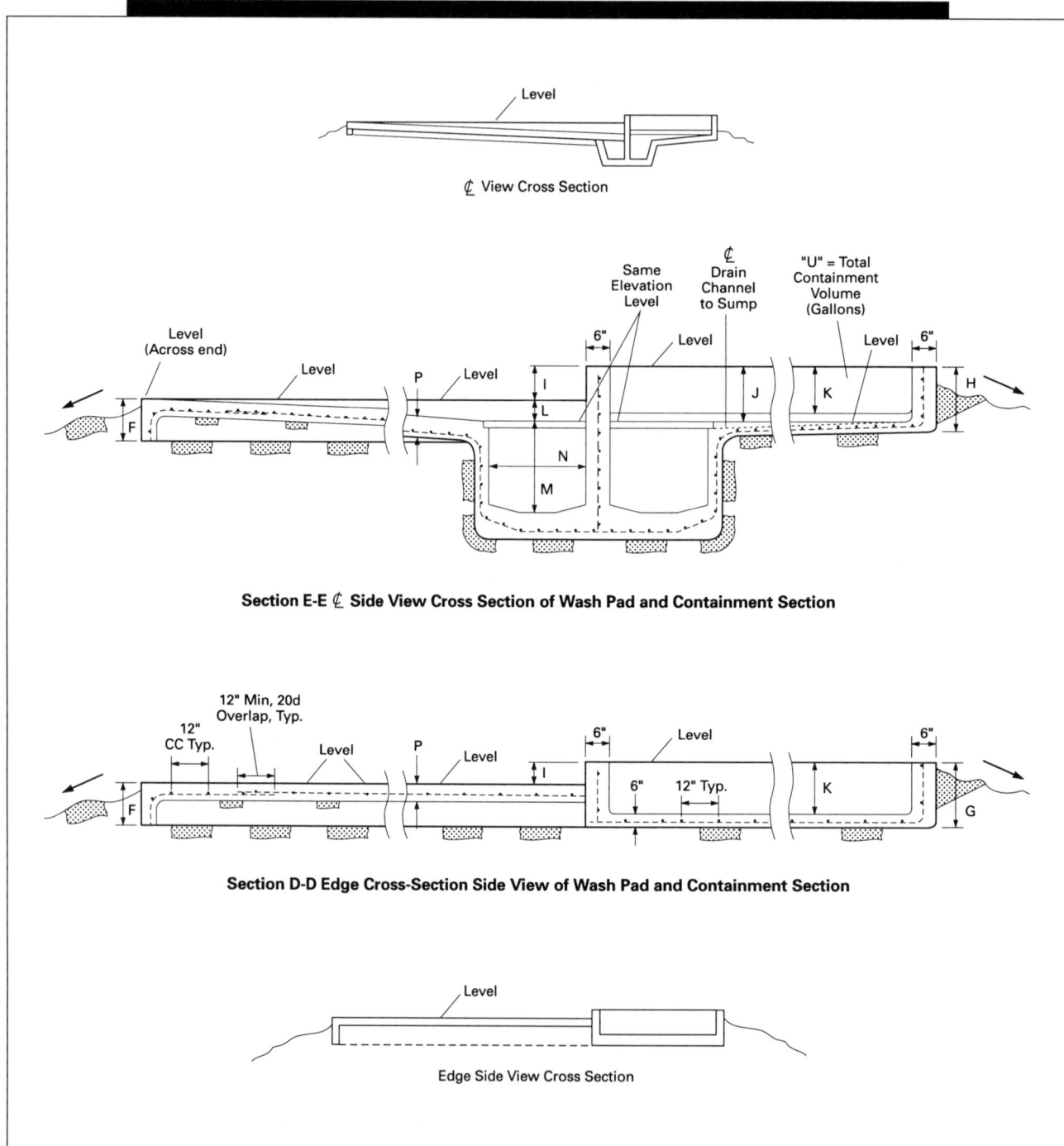

Level

₵ View Cross Section

Same Elevation Level

₵ Drain Channel to Sump

"U" = Total Containment Volume (Gallons)

Level (Across end)

Level

6"

Level

P

Level

I

L

J

K

6"

H

F

N

M

Section E-E ₵ Side View Cross Section of Wash Pad and Containment Section

12" Min, 20d Overlap, Typ.

12" CC Typ.

Level

P

Level

6"

Level

6"

I

6"

12" Typ.

K

F

G

Section D-D Edge Cross-Section Side View of Wash Pad and Containment Section

Level

Edge Side View Cross Section

Side Elevation Cross-Section Views of Wash Pad with Straight Sump Walls

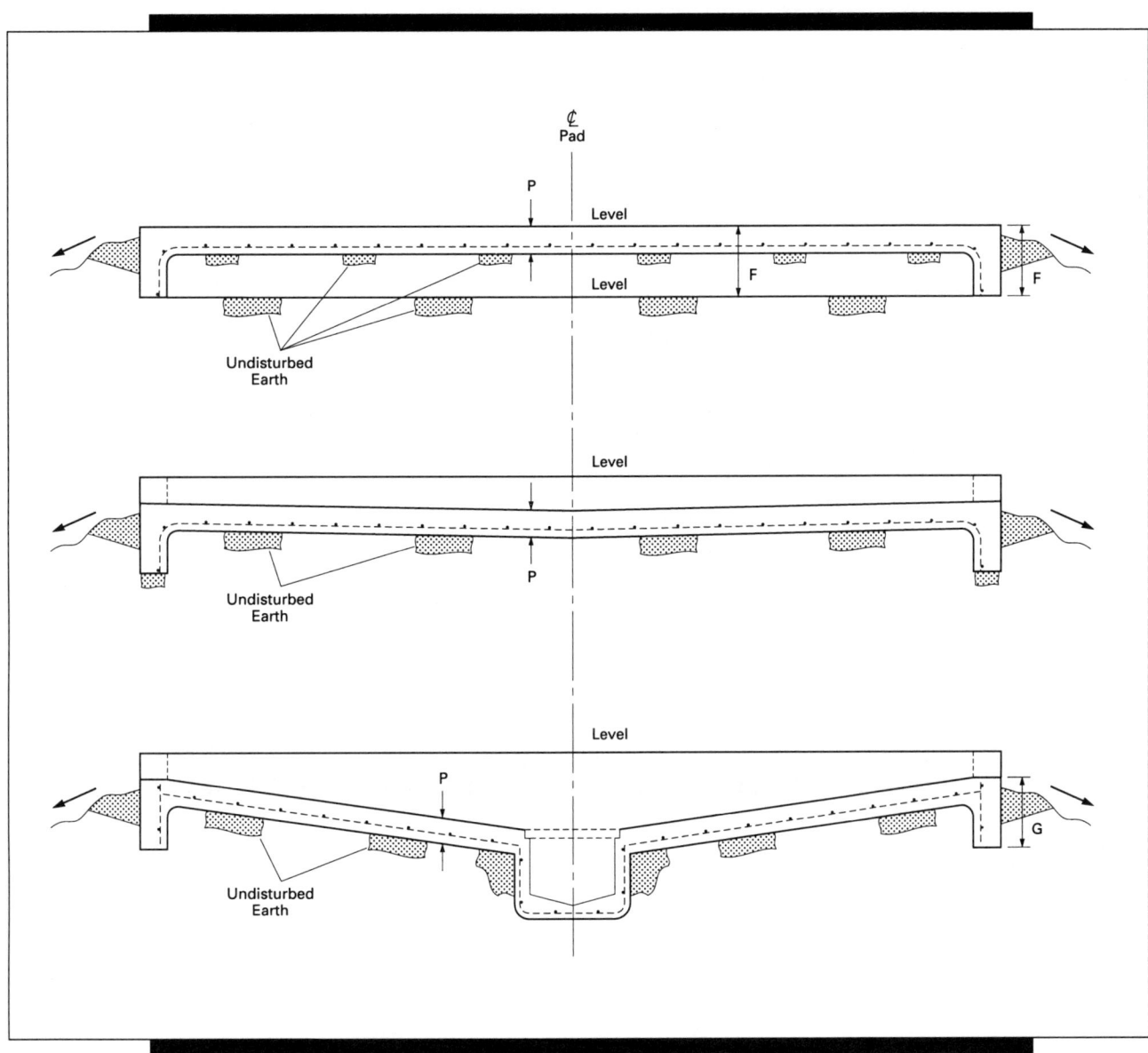

Cross-Section End Views of Wash Pad at Three Locations

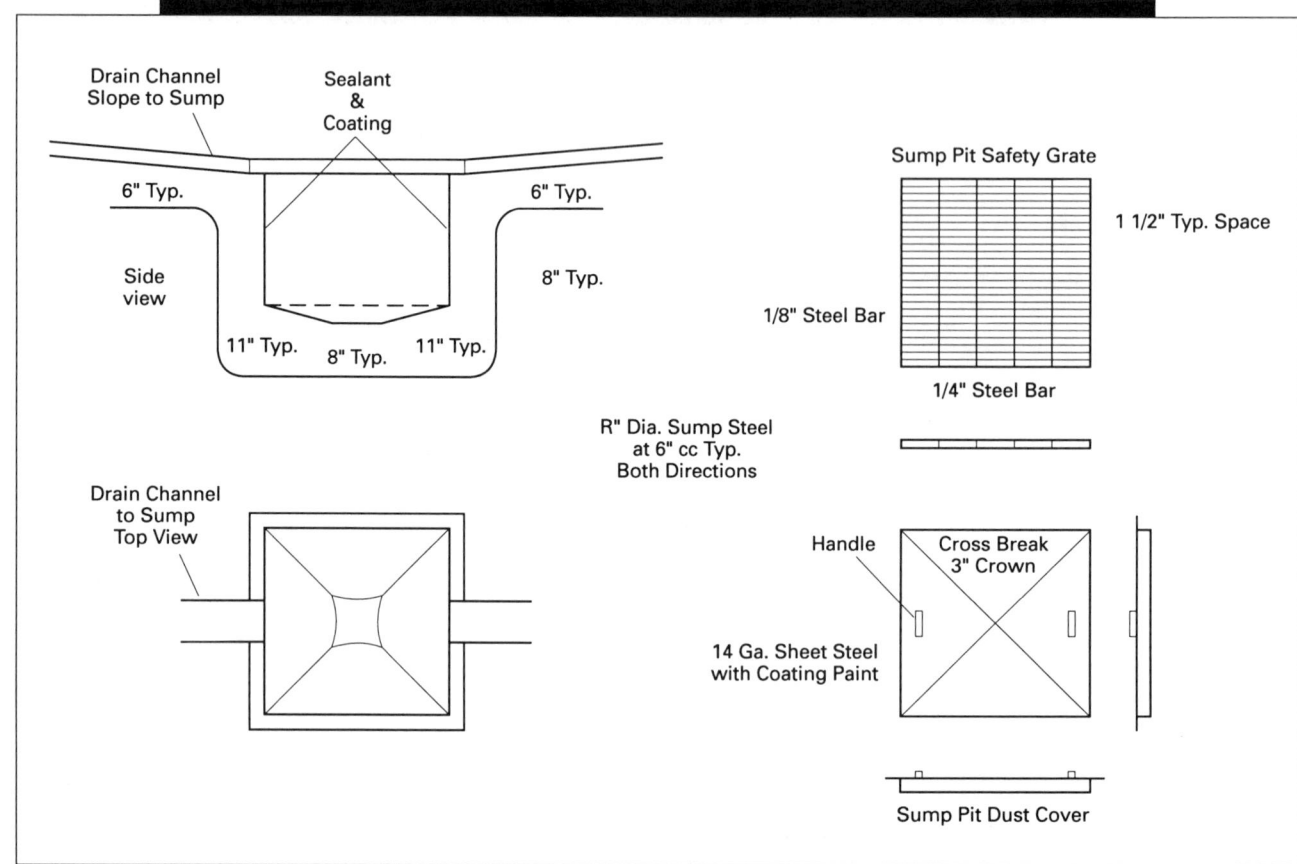

Cross Section and Detail of Self-Contained Sump

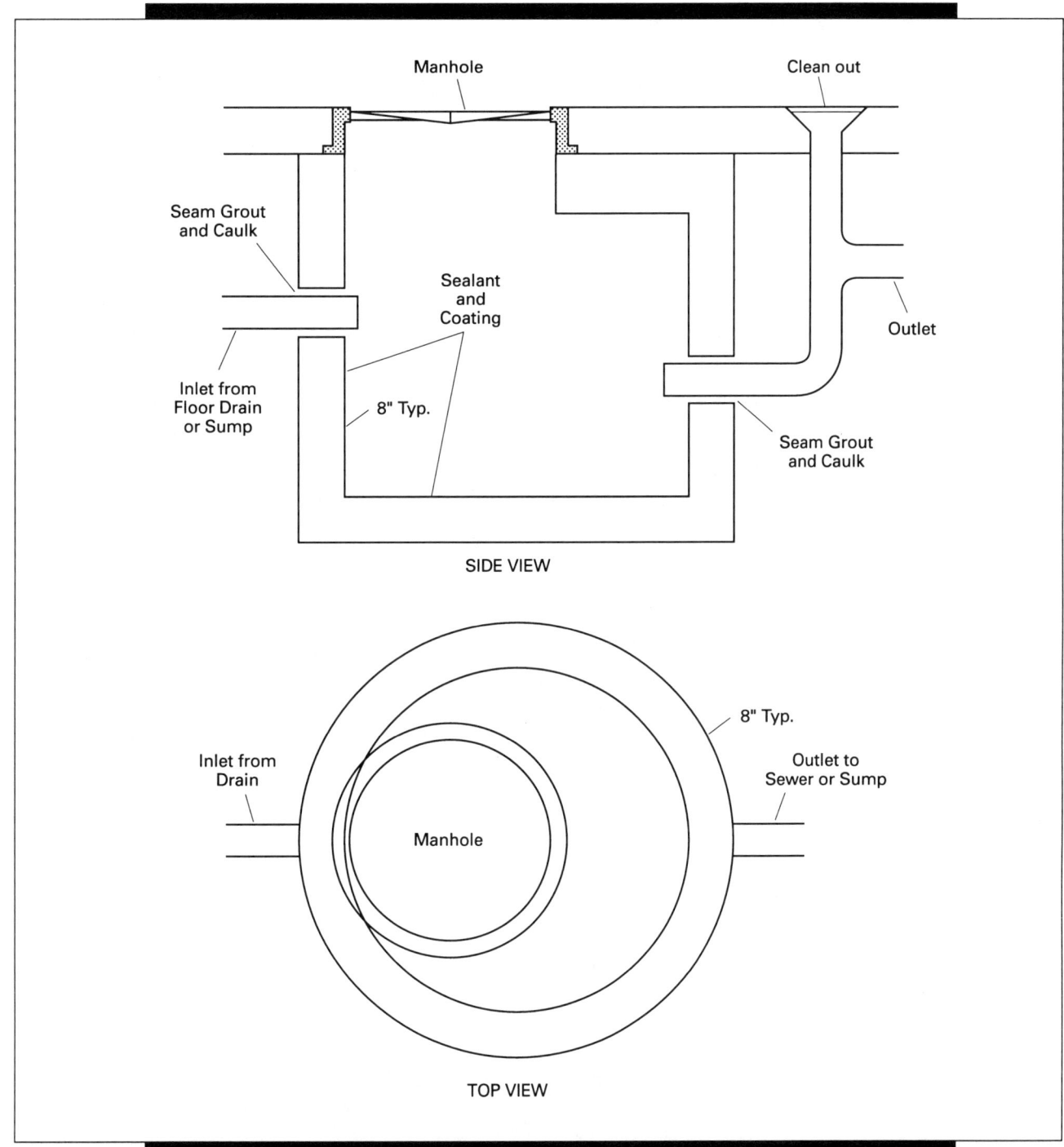

Cross Section of Sand, Mud, and Oil Interceptor

Questions and Answers

Repair and/or Lining Secondary Containment Areas

Q. What is the minimum protection required for secondary containment?

A. This may vary from state to state. Check with the person in charge of inspection in the appropriate state (e.g., Chief of Department of Agriculture, Hazardous Materials Site Inspector). Basically, the requirement is for a crack-free surface that will not allow penetration through the structure into the earth by the liquid that is contained in the primary vessel. To meet this minimum requirement means repairing any cracks found in the structure as quickly as possible (24–48 hours after discovery is prudent). This would also include expansion joints and wall-to-floor joints as well as lining, if applicable. **IMPORTANT!** These repair materials must be resistant to immersion in the liquids stored in the tank(s) for at least three (3) days.

Q. What is recommended for random crack repair, assuming such cracks are hairline, or small, and are "non-working" or "non-moving"?

A. For easy, fast, and long-lasting repair, the use of an epoxy-fiberglass crack repair kit is suggested. A 6-inch-wide fiberglass tape laid up with epoxy has 10,000 pounds tensile strength (concrete has about 300 pounds tensile strength). To assure a tough, chemical- resistant bond that will handle most chemicals, a solid epoxy mastic is recommended so repair can be made to vertical as well as horizontal cracks. Full instructions and all tools necessary for installation are furnished with most kits. Unskilled labor can make the repairs quickly and easily. One or more of these kits should be kept on site for emergency repair.

Q. What repair material is required for expansion joints or moving cracks?

A. For exterior or unheated containment areas subject to freeze-thaw conditions, or where expansion-contraction may exceed 25% elongation, a polysulfide sealant is recommended. Use of non-sag grade is required for walls. For application not subject to freeze-thaw conditions, a trowelable flexible grade filler may be used for easier and less expensive installation.

Note: Avoid silicone or other caulking, as it is unsatisfactory for this service. Only materials designated for immersion service should be used!

Q. If the walls of a containment area are constructed of Haydite cinder or concrete blocks, or concrete walls that have "honeycomb" or "bug holes," how does one eliminate them?

A. If block is very open or concrete is honeycombed, plastering the wall with an acrylic polymer concrete up to ⅛" thick is usually the best answer. If chemicals stored in the area are corrosive to concrete, a suitable sealer or lining may also need to be installed. If walls are tight concrete block or concrete with some bug holes, a trowelable epoxy filler may applied by squeegee or scrape troweling. Several applications may be required to fill all holes. Imperfections may be sanded off and sealer or lining may be applied over the filler. If a sealer or lining is to be used, filler is preferable to polymer concrete for this application because of it's chemical resistance and bond strength, as well as length of cure time. Repairing with a polymer concrete requires a minimum of seven days' cure time (28 days is best) before topcoating.

Note #1: A trowelable epoxy filler is highly recommended for coving wall to floor in all instances, as this is usually a weak spot subject to seepage.

Note #2: Repair and fill a small "mockup" area first for owner's or operator's approval before proceeding. Some owners, operators, or inspectors may want the area to look like a "living room wall," and that procedure can become expensive.

Q. What is the best way to repair holes, spalls, or deep patches in walls and floors of containment areas?

A. For the most chemical-resistant and quickest repair, or when sealers or liners are to be used, a trowelable epoxy filler is recommended. Trowelable grade filler may be used for horizontal or vertical repairs. If repair is over 1" deep, apply in layers to avoid excess heat buildup. If patches exceed 2" deep, polymer concrete is recommended. If the area does not require a sealer or liner, then this may be the best choice.

Q. How does one repair walls where active water is seeping or running through the cracks?

A. The answer to this may vary with the particular situation. If seepage is severe and continual throughout the seasons, an exterior drainage system may have to be installed. **If the floor heaves and hydrostatic pressures are involved, the floor cannot be sealed or lined successfully until proper drainage is engineered! Consult a qualified engineer to solve the problem!** If seepage is localized and occurs occasionally, then repair will usually be successful. To make such a repair on walls, first use a heavy-duty drill with a carbide bit to bore ½" or larger holes along the base of the wall and allow water to drain. Then repair cracks down to weep holes using a fiberglass epoxy crack repair kit and allow to cure for at least four days. Then fill the holes with hydrostatic cement to about ¹⁄₁₆" from the surface to stop active water. When dry, fill and fiberglass using fiberglass crack repair kit.

Note: It is suggested that trowelable epoxy fillers be used to cove wall-to-floor joints and corners before filling weep holes.

Q. How does one determine what lining to use for a secondary containment area?

A. The answer to this depends on many variables. This manual lists linings that are suitable for various uses and supplies chemical-resistance data. This data shows resistance to *chemically pure* substances; if mixtures of chemicals are involved, or chemicals for neutralizing spills, they must be checked. By reviewing this data, it should be possible to determine, in most instances, the system that one wants to use. Some of these systems are "user friendly" and easy to install on a do-it-yourself basis. Other systems are designed for very corrosive hazardous chemicals and may require fiberglass cloth reinforcement or expert trowel application over large areas. These systems should be installed by experienced professionals. An engineering service will assist in choosing the proper system that will be most cost effective. In order to do this, one should furnish the engineer with complete information on the installation as follows: Size; age; construction details and materials; blueprints; materials stored; ground moisture conditions; present condition; photos; location; presence of a vapor barrier installed below slab; previous coatings or linings and condition of same; installation and shut-down time; exterior, interior, or truck loading/unloading area; slip resistance, if required; traffic, if any; and type in secondary containment area. Will in-plant personnel or professional applicators be hired for installation? Furnish any other information to the engineer that may be helpful.

Dikes

Q. Why should it be required that boxes, jugs, and bags not be stored on the dike floor?

A. The containers may be contaminated with product. The dike should be clean and free of contaminants.

Q. How does one seal cracks in the dike?

A. Contact the equipment supplier for the chemical-resistant caulking. The joint at the floor and wall should be sealed with the caulking recommended by the equipment supplier.

Q. What is meant by a permanently sealed valve in the dike wall?

A. The valve is removed and sealed with concrete if the dike is constructed of concrete or concrete block.

Q. Is a steel dike acceptable?

A. Yes, if it is approved by the state and constructed of ⅜" plate steel, welded at all seams. Stock tanks, used fertilizer tanks that have been cut down, or steel less than ⅜" steel is not approved for storage of pesticide products.

Q. Why are poly dikes not approved?

A. Poly dikes are subject to cracking after prolonged exposure to sunlight. The pump in most cases is stored outside of the poly dike, causing potential spills and other problems.

Q. What type of construction should be used to avoid seams at the dike wall and floor?

A. Use a floating dike in which the dike and wall are poured together.

Q. If a concrete block wall is constructed, should it be filled?

A. Yes, the blocks should be filled with re-rods spaced every 18 inches on center, and the re-rods must be attached to the floor of the dike. This will tie the floor and the wall together. Concrete block (cinder block) walls must be capped to avoid being filled with water, freezing, and cracking. Furthermore, an approved coating must be applied to seal the pores.

Q. How many man-hours should it take to pour a cubic yard of concrete?

A. The average time for concrete construction is 2 to 2.3 man-hours per cubic yard of concrete.

Groundwater/Surface Water

Q. How does one dispose of surface water?

A. The best way to eliminate surface water is to construct a roof over the site or put the entire bulk site in an enclosed building.

Q. If a building is not feasible, what action should one take?

A. Keep the facility clean and free of chemical, and pump the water in a holding tank to be sprayed on a field at a convenient time.

Q. How does one forecast the amount of water that will be handled from the bulk facility after a rainfall?

A. Every inch of rainfall equals 625 gallons of water per 1,000 square feet.

Rinsate

Q. What should be done with rinsate water?

A. Use the water in tank mixture and spray the water on the labeled product crop on the next fill. Do *not* store the water from the rinsate system or the application truck for a long period of time. It should be reused daily.

Q. Should one use a drive-up ramp to the fill/rinsate pad?

A. Yes, at least the length of one revolution of the tire.

Q. What capacity should the rinsate system have?

A. The rinsate system should hold 110 percent of the largest tank that is filled on the pad to allow total protection for the dealer. You should construct the system for your trucks or spray rigs.

Pumps and Meters

Q. How does one winterize pumps and meters?

A. Most chemicals should be drained from the pump; therefore, check with the chemical supplier. If the chemical must be drained from the pump, there should be antifreeze in the system.

Q. How does one test the meter for accuracy?

A. With a five-gallon test bucket, not a scale.

Tanks

Q. Can a glass-lined beer tank be used?

A. Yes.

Q. If a poly tank has been inside for eight years, is it safe to use?

A. Yes. Sunlight destroys poly tanks over years; a tank in a building has a much longer life than one out in the sunlight.

Q. Should one recommend only stainless?

A. No. Poly tanks, if constructed properly, have a long life, are economical, and offer the same benefits as stainless. Stainless is the highest-quality tank.

Q. Why are bottom-filled tanks recommended?

A. For safety to the area and the driver.

Summary Discussion

The modular chemical-handling facility described here allows integration of the major management functions of commercial pesticide applicators. These functions include hauling, mixing, transfer, loading, storage, and security of concentrate or finish mix pesticides; pesticide rinsate storage, handling, and reuse as makeup water; and the washing, loading, rinsing, clean-out, and parking of application vehicles and equipment.

The size or specific type of operation does not matter; the system by its modular concept can be downsized, up-sized, or arranged to fit essentially any pesticide- and/or fertilizer-handling entity.

The purpose of this guide is to provide a comprehensive review of design and construction concepts that should provide a suitable, state-of-the-art chemical handling and management facility. The modular flexibility allows these design principles to be incorporated with modifications to meet specific needs in most states in the United States.

With new legislation in several states and the EPA's focus on point-source containment sites, it becomes increasingly important for the pesticide-using industry to implement the basic principles covered in this guide.

APPENDIX I

Equipment Standards

Tanks—General

Construction Materials

Stainless steel is the most versatile and desirable material for a tank. It will provide long life, low maintenance, and increased safety, but it initially costs more than other materials.

Polyethylene tanks can be used where state and/or local regulations permit. They are low in cost, easy to maintain, versatile, and readily available, but they have a limited life.

Filling Port

Bottom loading into a storage tank is recommended both for safety and to prevent air entrapment. The bottom outlet should have a stainless steel valve with a locking handle and a tee to provide a filling port, which should be fitted with a valve and male Kamlok fitting.

Venting

Free air movement in a bulk tank can cause a variety of problems, such as "skinning" of suspensions, contamination, or evaporation. To prevent these problems, all bulk tanks should be fitted with a conservation vent that opens and closes within the designed pressure limits of the tank.

Vent capacity should be large enough to permit unloading from a tanker equipped for high-volume discharge. In no case should the vent permit water to enter the tank.

Size

To facilitate filling and expansion of the tank contents, provide a minimum of 5 percent excess capacity (head space).

Gauges

Sight gauges are not recommended for bulk chemical tanks both for security reasons and because of their incompatibility with farm chemicals. If a site gauge is used, it must contain a deadman valve. Float gauges are safe by their design. A meter can also provide inventory control.

Scales

A quality platform scale, graduated in no more than ½-pound increments, can be an accurate way of filling mini-bulk tanks.

Security

A fenced area or enclosed building is desirable. In addition, positive-locking stainless steel ball valves must be installed at the loading and discharge point and should be kept locked. To protect against vandalism, weld the nut on top of the handle to the valve shaft. This prevents removal of the handle. A locking ball valve at the tank outlet is always required.

Sampling

If sampling is required by regulation in an area on a demand basis, one can sample at the discharge nozzle of the system. Taking a sample and labeling it at the initial delivery helps keep a representative sample on hand.

Recirculation and Bypass

One should be able to mix the tank contents completely by recirculation. This can be most easily accomplished by using an in-tank eductor. The entire tank volume should be recirculated three times before it is removed. This eliminates the effects of settling and ensures a more uniform product. The inlet of the recirculation line should be below the liquid level to prevent introducing air into the product. Do not air sparge.

Flowables should be thoroughly recirculated (at least three tank volumes) one week prior to the use season. Flowable herbicides should be recirculated for three tank volumes weekly during the use season. When loading out, open the recirculation valve slightly to allow some bypassed product to agitate the tank contents.

Manhole

A manhole on top of the tank is recommended for inspection of the interior of the tank only. At no time should anyone enter the tank to clean it. The manhole must have a gasket to ensure a good seal and to prevent contamination, especially by rainwater.

Storage Tank Cleaning

Tanks must be designed for easy cleaning. High pressure or steam and a means to add detergent should be incorporated into the design. Tanks must be thoroughly dried after cleaning and before refilling.

Before product is placed in a tank, the tank must be free of all contaminants, such as organic solvents, petroleum products, acids, fertilizers, and other pesticides, and free of water.

Workers should not be allowed to enter a storage tank that has been used to store pesticides.

Protection for workers cleaning a tank that has been used to store a product must be based on the physical, chemical, and toxicological properties of the product.

It is best to clean a tank immediately after emptying. Dried product is much more difficult to remove, and other methods of cleaning may be necessary.

Pumps

The pump should be located as close to the tank as possible. A suction line longer than six feet should be avoided. Pumps push better than they pull.

Type

Centrifugal pumps are usually satisfactory. More viscous products may require positive displacement pumps equipped with safety devices, such as a relief valve.

Capacity

The size of pump needed depends on the length and diameter of the line, the required flow rate and lift, the amount of bypass, and the viscosity of the product. Consult a pump supplier with this information and ask for a recommendation.

Housings

Carbon steel or stainless steel is preferred.

Seals, Gaskets, and O-Rings

Only viton or teflon seals and gaskets are recommended. O-rings should be of viton. Use only teflon for handling these materials.

Installation

Install pumps so that they can be removed easily for servicing and winter storage. Provide clearance for the pump drain plug so that product or rinse water in the lines can be drained and disposed of properly.

Maintenance

Keep spare seals, gaskets, and O-rings on hand. Store the pump under cover in the winter, and winterize with an antifreeze-and-water mix, much as one would prepare a car radiator for winter.

Motors

Types

Always use totally enclosed fan-cooled electric motors. Gasoline engines should be used only where electric power is not available and used outside a building. Electric motors are also preferred because they can be operated remotely.

Size

Motors should be sized to fit the required pump performance. Be sure the motor is adequate for recirculation.

Wiring

The most efficient motors are 220-volt three-phase. Be sure the wire gauge is adequate for the distance and load requirements.

Protection

Protect all electrical components from weather and chemicals, especially liquid fertilizers, by installing control boxes remotely and under a roof.

Meters

The meter is the heart of any bulk system. Only the best quality should be installed. Consider the extra features, many of which can aid in inventory control and can be connected to in-house computers.

Product Measurement

Use a separate meter for each tank. Variations in viscosity and specific gravity between products make this essential.

Locate the meter close to the discharge point for ease of operation and for better control. Meters are not accurate outside their specified range. Control product flow accordingly. Most meters are made to operate between five and thirty gallons per minute.

Type

The Tokheim 680 series is the industry standard. It is highly accurate with a piston-type design that can measure only volume and is not affected by viscosity changes. The flow rate for this series is 0 to 40 gpm.

Another popular meter is the LC. It has a rotary design, is $1\frac{1}{2}$ inches in size, and has a flow range of 5 to 60 gpm.

Size

Each meter has an optimum flow-rate range at which it is most accurate. To measure effectively, it must be run within its range. Excess flow from the pump should be diverted through a bypass. The meter should be matched to the flow rate of the pump.

Calibration

A meter should be calibrated for accuracy before being put into operation each year and should be recalibrated weekly during the season, especially as product temperature changes. To calibrate, use a five-gallon test can. This is a specially designed vessel available from meter vendors.

Procedure for Calibrating a Meter for Product Resale

1. Recirculate the tank contents at least once before calibrating.
2. Pump a small amount of product through the delivery hose to eliminate any air pockets between the meter and nozzle. Calibration should be done with the delivery hose full and the meter set at zero.

3. Pump product into the test can. Stop when the meter reads 5.0 gallons. Compare the meter reading with the level in the can. The test can is marked in cubic-inch increments of which there are 231 per gallon.

4. Adjust the meter up and down as needed and repeat steps 2 and 3.

5. Once the test can and the meter agree, repeat steps 2 and 3 again to ensure repeatability.

6. Check the calibration at least once per week, because changes in product temperature and equipment wear can affect meter accuracy.

Plumbing and Fittings

Layout

When designing a system, keep the lines as short as possible and use a minimum amount of fittings. Use a line size that is adequate to supply the pump. Ensure that there is adequate support to prevent sagging.

Material

Use stainless steel pipe and fittings wherever possible to reduce corrosion. Do not use PVC or other materials that will be attacked by agrichemicals. Use a quality pipe thread compound on steel fittings. Do not use teflon tape on steel fittings.

Size

Size pipes in the system to achieve the desired flow rate. Check for compatibility with pump specifications.

Valves

Lockable stainless steel ball valves must be used on the tank. Poly tank valves are suitable for use in a system after the tank.

Gaskets

Use gasket materials that are resistant to the products used. Teflon or vitron is recommended.

Hoses

Cross-linked polyethylene-lined hose is best, although EPDM is suitable in most instances.

Screens

Do not install the strainers between the tank and the pump. However, a strainer between the pump and the meter is recommended.

Installation

Settling and shifting are likely to occur, so build in enough flexibility to prevent strain and possible rupture. Protect from accidental damage by designing the system so that the plumbing is out of the way of frequent human and vehicular traffic.

To protect water sources from contamination, plumb products separately from any source of water and use a back-flow valve to prevent back siphoning of product into water supplies.

Winterizing Equipment

After the application season, but before winter, be sure to clean, maintain, and winterize the bulk system.

Tanks

Inspect the conservation vents and valves. Check for signs of rust and corrosion; paint and repair as needed. Remove water from empty tanks and seal all places water could seep in.

Pumps/Meters

Flush out with clean water and fill with antifreeze and water mix to prevent rusting. Replace seals, gaskets, and O-rings as needed and store indoors if possible.

Plumbing

Check the overall condition. Flush with clean water and fill with antifreeze-and-water mix. Be sure to flush the antifreeze from the system and drain and dry before using again.

Flexibles

Check the condition, flush with clean water, drain, and store under cover.

Site Standards

Dikes

General

Store pesticides only in facilities equipped with a dike made of poured concrete, concrete block (cinder block), or steel. Do not store pesticides in any facility that is undiked or has earthen, gravel, or poly dikes.

The dike may not contain any outlets. If a relief valve or outlet is present, it must be permanently closed, plugged, or sealed with concrete or steel. This restriction is designed to prevent the accidental loss of product from an open valve or outlet that renders the dike useless. A sump pump with a locked switch, not automatic, is recommended for the removal of water that may accumulate due to rain or snow.

The dike may not be used to collect or hold rinsate water. Instead, a separate holding or collection tank should be used for rinsate water. The dike should be above the surrounding topography at the site to permit an inspection of the dike's integrity. The ground around the dike should be graded to carry surface water away from the dike. Unless stricter state standards apply, the dike must have sufficient capacity to contain a minimum of 110 percent of the contents of the largest tank and the displaced volume of the butts of the tanks contained within the dike.

The specifications listed below must be followed to meet the standards for bulk sites.

Concrete Dikes

Most states require the use of concrete rated at a minimum of 3,500 pounds psi for poured dikes, with re-bar placed on no greater than one-foot centers. Walls and the dike floor should be a minimum of six inches thick. The dike floor should slope slightly to one corner to permit collection of water by a sump pump.

The walls and floor of the dike should be poured together. If the walls and floor of the dike are poured separately, a chemical-resistant caulking compound should be used to seal the seam.

Construction of Concrete-Block or Cinder-Block Dikes

It is recommended that the floor be poured first. Construct the dike by laying concrete blocks on top of the floor with re-bar reinforcement from the floor to the block wall. Re-bar should be placed no greater than one foot on center.

For concrete-block dikes constructed outdoors, most states require that the walls be filled and capped to prevent winter water damage to the walls. Interior dike walls must also be capped to facilitate maintenance and cleaning. If a crack develops, use a chemical-resistant caulking compound to repair the dike.

Steel Dikes

Steel dikes must be constructed with a minimum of $\frac{3}{8}$ inches of steel welded to prevent leaks at all seams. Beams should be welded to the floor under the containment dike to allow visual monitoring and inspection of the dike. The dike should be installed on a crushed-gravel or concrete base.

Livestock tanks and old fertilizer tanks do not qualify as dikes for the storage of pesticides.

Poly Dikes

Poly dikes are not sufficient for diking at facilities storing pesticides.

Rinsate System

The rinsate system must be constructed of concrete and must cover at least 500 square feet. The containment walls must contain all spills while filling or unloading trucks or mini-bulk containers. No water containing chemicals may be stored in the rinsate pad area. An aboveground holding tank system must be in place to recycle wastewater in subsequent applications.

Unless stricter standards apply, most states require that the rinsate system be constructed of concrete at a minimum of six inches thick, with re-bar placed on no greater than one-foot centers. The concrete should be rated at a minimum of 3,500 psi. The rinsate system should be attached to the dike, with the area covered by a continuous concrete floor.

The system should be designed with sloping sides to form a collection basin for liquid wastes. The approach ramp to the area should be the length of one revolution of the tires on the largest piece of application equipment. This will minimize the amount of dirt and gravel entering the collection system.

This area is not a dump site for leftover loads. It should be used for controlling accidental spills or releases from application or nurse equipment.

Security

Security measures include fencing or enclosing the bulk storage area in a secured building. Fences should be at least six feet tall around the dike area, and all gates must be locked when the facility is not in use. A secured building may be constructed around the bulk facility so that doors can be locked when not in use. The added benefit of a roofed or enclosed facility is the ability to manage precipitation and rinsate waters.

Sealants and Coatings

All dikes, floors, and wash pads should be treated with watertight, wear-resistant materials that are also resistant to chemical corrosion.

All expansion joints, seams, or cracks should be sealed with watertight, chemical-resistant caulking compound.

Cinder-block dikes or dikes with floors and walls poured separately should have a bonding agent applied to establish integrity; and a watertight, chemical-resistant caulking compound should be applied to the seams.

Inspection of Concrete Surfaces Prior to Application of Coatings on Concrete Floors

General

Concrete prepared with the same raw materials can vary from one floor to the next. This variance can be caused by discontinuities in cement, sand, and gravel, non-uniform mixing, the degree of vibration, and, particularly, the weather conditions during the pouring. Concrete poured in cold weather may not properly hydrate resulting in a weakened, powdery concrete. Tapping the floor with a hammer will give a dead sound over weakened concrete that requires repair. There can be physical problems, such as honeycomb, protrusions, tie rod holes, and form joints. When concrete is trowel-finished, a certain amount of cement leaches to the surface. This is called laitance and is quite weak and needs to be removed. In order to be coatable, a concrete surface must contain 6% moisture or less, preferably less. Hydrostatic pressure or the build-up of ground moisture can force water through concrete surfaces, blistering enamel-type finishes. In order to gain good coating adhesion, the pressure of ground moisture must be relieved either by the installation of drain tile or, during construction, the installation of appropriate vapor barriers. The chart shown below gives the appropriate curing times for new concrete.

Concrete Curing

Temp °F.	Days
40	56
50	41
60	32
70	28

Construction chemicals can form intervening barriers. These include curing and sealing compounds, waxes or release agents used in tilt-up construction, oil and grease spillage, and chemical contamination from acids or alkali compounds spilling on the uncoated surface. Aggregate contaminants such as lignite are often found in river gravel and can make concrete virtually uncoatable. Floor hardeners can also affect coating adhesion. The best floor is one that is cured to a compressive strength of a minimum of 3800 psi and cured mechanically with either plastic or burlap.

In order to get the history of a concrete floor one needs to ask the following questions: a) What was the curing time allowed for the concrete and at what temperature? Was the concrete laid over a vapor barrier? Is the location of the concrete in an area that could present moisture problems or hydrostatic pressure? b) If previously coated, what was the system? c) If previously coated, what is the condition of the existing coating system? d) If failing, what kind of failure is occurring? Blistering? Peeling? Dissolution? e) Take photographs (if photographs are allowed), carefully mark them, and send them to Steelcote with your comments.

Moisture Test

There are two methods of testing for moisture:

Method A
*ASTMD-4263 Standard Test Method for Indicating Moisture
in Concrete by the Plastic Sheet Method*
This method suggests that for every 2500 square feet an 18″ x 18″ piece of four-mil clear plastic sheet be taped to the surface using duct tape 2″ wide and 50 millimeters thick. The sheet must be taped completely on all sides to prevent air leaks and remain in place for at least 16 hours. After that time, it should be removed and the underside checked for moisture. If after 16 hours no droplets are observed on the inside of the sheet, the concrete may be coated after proper surface preparation. If droplets are observed, the coating application cannot take place until further drying has occurred. The test should then be repeated prior to any coating work.

Method B
A Delmharst (or Equivalent) Moisture Meter
By drilling small holes into the concrete surface, copper nails can be inserted at the width of the prongs on the moisture meter. A direct reading may then be taken and recorded on the meter gauge. This method is more suitable for plaster, but may be used to augment Method A.

Intervening Barriers

This test can be performed rather easily by simply putting droplets of water on the surface with a plastic spray bottle. If the water beads up, there is an intervening barrier. If it soaks into the concrete, there is no intervening barrier. A further check can be made using a concrete clean-and-etch material to see if the material will react by bubbling or effervescing with the concrete to form an etch pattern, as well as produce a bleached surface. The etching solution must also remove any laitance and produce a surface resembling medium-grade sand paper. If no reaction takes place, it means there is an intervening barrier that must be treated or removed.

Old Coatings and Sealers

This section is also partially covered above. Many older coatings slowly but surely deteriorate by the very presence of the concrete. Adhesion tests should be run with the new coating system, if the old coating cannot be removed. Testing should include at least two coats of the recommended finish system with a proper tie coat. If in doubt, removal of the old coating is recommended.

Mock-up

To ensure a successful floor installation, it is necessary that a 400-square-foot trial area be installed on any job that has a large area to be coated. This area should include the surface preparation. There are several reasons for the use of a mock-up: 1) It will show everyone exactly what the floor system is going to look like when the job is finished. 2) The surfaces can be checked for adhesion in accordance with ASTMD-3359 Method B. Method B consists of scoring the mock-up with a razor blade all the way to the concrete surface itself. Create a 1" square with cuts ⅛" apart and perpendicular to one another so that there are approximately 64 small squares within the 1" square. Press scotch tape or some other pressure-sensitive tape onto the square very thoroughly using a pencil eraser; then quickly pull it off. It must show a minimum of 98% adhesion unless, of course, concrete comes away with the tape. If this happens, it suggests incomplete surface preparation. 3) The mock-up will indicate the amount of material required to do the job and will help in giving an accurate estimate of the material requirements.

Once the owner's or engineer's approval is obtained, the criteria are established for the balance of the installation. Since concrete is a variable, care needs to be taken in installation of the coating system to make sure that each segment of the floor comes out looking like the mock-up. Therefore, if convenient, the placement of the mock-up should be in an area where maximum repair might be necessary, if the owner desires the floor to be of uniform appearance. This will also give the applicator an opportunity to estimate the amount of material required to make any such repairs. While the mock-up is being installed, it's prudent to check the wet film thickness with a Microtest Wet Film Thickness Gauge or equivalent. By calculating the number of square feet within a square surrounded by cracker joints, material coverage rates can be controlled.

Surface Preparation for Coating a Floor or Repairing a Crack

A major contributor to the success of a floor coating or crack repair project is the surface preparation. It is important that the concrete surface be free of any lubricants, dust, grime, paint, or other materials that may affect the coating's ability to adhere to the concrete.

In several cases the concrete may be stained with a chemical or other material. In this instance, it is important to spend additional time insuring that the contaminants are removed, even though the concrete may remain stained.

Surface Preparation Method #1

Step 1
Power wash, using water and a strong detergent (e.g., industrial grade soap, Spic & Span®, Tide® detergent), the concrete surface to be coated and all cracks to be repaired.

Step 2
After power washing or rinsing the surface with clean water, use muratic acid full strength on the surface to be coated to etch the concrete . A mop may be used to spread the muratic acid. Allow acid to work for 10 minutes (it will "bubble" when working). Using a stiff push broom or brush, scrub the area, making sure to remove all contaminants on the concrete surface and clean any cracks that are present. When this is complete, power wash or rinse the area with water. If some areas still appear to be contaminated, repeat this step on those areas.

Step 3
Once the area has been power washed or rinsed with clean water, mix at a ratio of five gallons of water to one pint of 28% aqua ammonia. Once the material is mixed, pour the solution over the cleaned concrete. Using a mop, move the solution around for five minutes and let stand for approximately 15–20 minutes. This will neutralize the concrete from the acid material used in step 2.

Step 4
Rinse the area thoroughly with clean water and vacuum up or squeegee off all excess water. (It may be necessary to blow out the cracks with an air compressor.)

Step 5
Allow the area to dry *completely* before applying the coating or repairing any cracks.

Surface Preparation Method #2

Step 1
Sandblast the area to be coated, making sure to remove all contaminants from the concrete surface and clean any cracks that are present. *Depending on how contaminated the concrete is, it may be a good idea to power wash the area with a strong detergent before sandblasting.*

Step 2

Once sandblasting is complete, sweep or rinse the area with water, making sure to remove any sand particles remaining on the area to be coated. *If after sandblasting the concrete remains stained, repeat sandblasting operation. Acid etch and neutralize as mentioned in method #1 if necessary.*

Step 3

Allow the area to dry completely before applying the coating or repairing the crack.

Non-moving Crack and/or Seam Repair in a Corner

Step 1

After the surface is cleaned, apply a trowelable filler to the corner using a 1–1½″ putty knife.

Step 2

In order to obtain a smooth rounded corner, use a 1½″ PVC pipe. Wet the end of the pipe with water and run along the corner, smoothing the trowelable filler and creating the rounded cove.

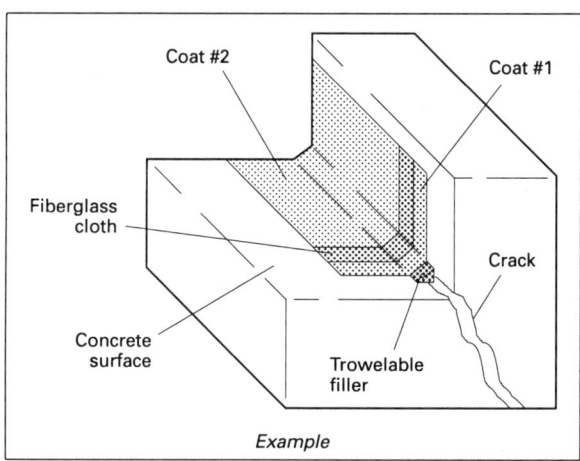

Example

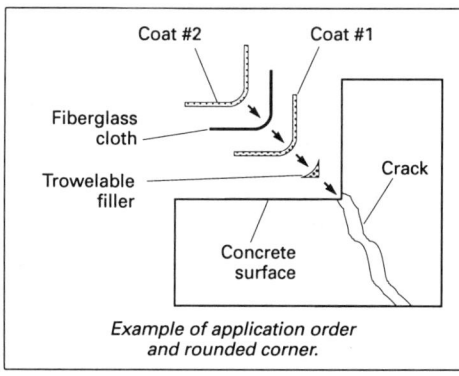

Example of application order and rounded corner.

Step 3

Mix a 3-gallon unit of sealant in a 5-gallon bucket, then pour that mixture into a second 5-gallon bucket. (Be sure to scrape the sides and bottom of the first bucket to remove all unmixed material.) Mix the contents in the second bucket for a minimum period of three minutes. (This will ensure a complete mix of the material and will not leave soft spots in the floor or wall coating.)

Step 4

Apply one coat of sealant using a nylon loop paint roller. The sealant must be applied within 16–24 hours of the trowelable filler application. Any additional time will require the trowelable filler to be sanded in order to obtain proper adhesion of the sealant. *A nylon loop paint roller may be purchased from a building supply store.*

Step 5
While the first coat of sealant is still wet, work in the fiberglass cloth with the roller until the sealant has soaked through the cloth, overlapping the cloth at all splices. *It is strongly suggested that the fiberglass cloth be cut to follow the crack in advance. It is important to work in the fiberglass cloth before the sealant "sets up."*

Step 6
After about four hours (at 75°F minimum), apply a second coat of sealant on top of the first coat and fiberglass. Allow 48 hours to dry. Final cure is expected in approximately seven days at 75°F.

Non-moving Crack Repair on a Flat Surface

Step 1
After the surface is cleaned, mix the 3-gallon unit of sealant in a 5-gallon bucket, then pour that mixture into a second 5-gallon bucket. (Be sure to scrape the sides and bottom of the first bucket to remove all unmixed material.) Mix the contents in the second bucket for a minimum period of three minutes. (This will ensure a complete mix of the material and will not leave soft spots in the floor or wall coating.)

Step 2
Apply one coat of sealant using a nylon loop paint roller. *A nylon loop paint roller may be purchased from a building supply store.*

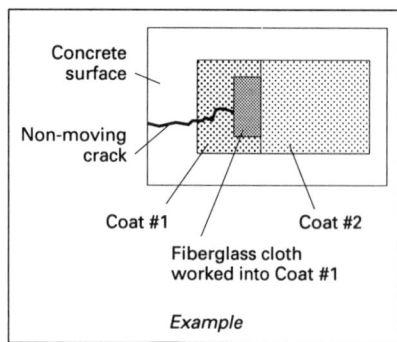

Concrete surface

Non-moving crack

Coat #1   Coat #2

Fiberglass cloth worked into Coat #1

Example

Step 3
While the first coat of sealant is still wet, work in the fiberglass cloth with the roller until the sealant has soaked through the cloth, overlapping the cloth at all splices. *It is strongly suggested that the fiberglass cloth be cut to follow the crack in advance. It is important to work in the fiberglass cloth before the sealant "sets up."*

Step 4
After about four hours (at 75°F minimum), apply a second coat of sealant on top of the first coat and fiberglass. Allow 48 hours to dry. Final cure is expected in approximately seven days at 75°F.

Additional step for non-moving cracks larger than ¼" in width:
If cracks are larger than ¼" in width, apply trowelable filler with a putty knife before proceeding with the steps listed above. See the example below.

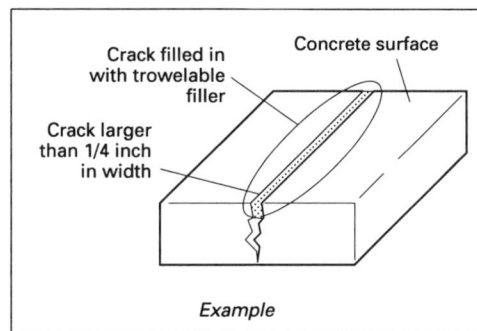

Example

When filling in a crack with trowelable filler, be sure to proceed with steps 1 through 4 above within 16–24 hours of application. Any additional time will require the trowelable filler to be sanded in order to obtain proper adhesion of the sealant.

Moving Crack Repair on a Flat Surface

Repair of moving cracks smaller than ¼" in width:

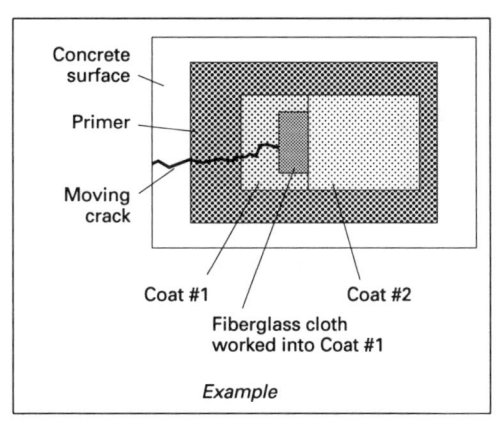

Example

Step 1
After cleaning the surface, apply a coat of primer to the surface of the concrete using an ordinary paint brush and allow to tack up (30–45 min.).

Step 2
After the primer has tacked up, apply one coat of caulk to the concrete surface and in the crack, if large enough.

Step 3
While the first coat of caulk is still wet, work in the fiberglass cloth with a roller until the caulk has soaked through the cloth, overlapping the cloth at all splices. *It is strongly suggested that the fiberglass cloth be cut to follow the crack in advance. It is important to work in the fiberglass cloth before the caulk "sets up."*

Step 4

After about 30 minutes (at 75°F minimum), apply a second coat of caulk on top of the first coat and fiberglass. Final cure is expected in 36–48 hours at 75°F.

Repair of moving cracks larger than ¼" in width:

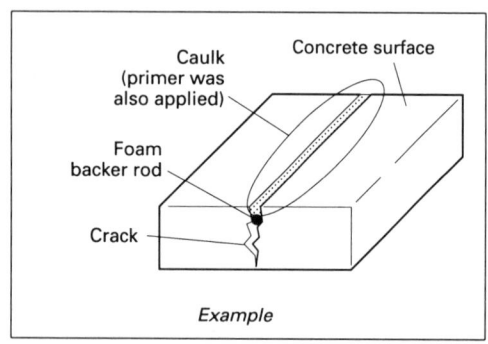

Example

Step 1

After cleaning the surface and crack, insert a foam backer rod into the crack. Insert only a quarter to a half inch, since this will serve as a filler and allow the caulk to expand and contract with ease, without pulling off the concrete.

Step 2

Apply a coat of primer in the crack and on part of the concrete surface using an ordinary paint brush. Allow primer to tack up (30–45 min.).

Step 3

After the primer has tacked up, pour the caulk into the crack.

Step 4

Final cure is expected in 36–48 hours at 75°F.

Epoxy Fiberglass Crack Repair Kit

Instructions for Use

READ ALL INSTRUCTIONS COMPLETELY BEFORE USING

SAFETY: Avoid eye or skin contact with the epoxy resin, hardener, clean-and-etch material, and ammonia. In case of eye contact, flush with water for 15 minutes. If discomfort persists, seek medical attention. Provide good ventilation, particularly when applying inside buildings. This is particularly important when etching and neutralizing the surface. If using inside tanks and other enclosed areas, wear appropriate, properly fitted air supplied respiratory equipment, NIOSH or MSHA approved (see label), during and after application unless air monitoring demonstrates vapor mist levels are below applicable limits. Follow respirator manufacturer's directions for respirator use. Follow other "Warning" information on the labels of the containers. If epoxy is sanded after curing, wear dust mask. Wear latex gloves when handling the clean-and-etch material, neutralizer, and epoxy. Wear protective clothing to avoid contact with the skin. For removal of the uncured material from skin and other areas, use mild soap and warm water. Use a hand cream containing lanolin to restore oil in skin after washing. Avoid using solvents for removing material from skin. Nail polish remover or lacquer thinner will remove the epoxy, before hardening, from clothing.

ENVIRONMENTAL: Note that ammonia water solution is used to neutralize acid etch to acceptable limits for disposal in sewers. Once epoxy has been thoroughly mixed and allowed to harden, it becomes inert and may be disposed of in a normal landfill. It is recommended that mixing paddles, etc., be thoroughly rubbed together after use so that any material remaining thereon is reacted. Once the material has hardened, the mixing paddles cannot be pulled apart without breaking and may be disposed of in a landfill. Any remaining material in the cans, if properly mixed together (Part A and Part B), will promote materials in both cans to harden. Once the material has hardened, it may be disposed of in a landfill. However, any material not reacted will have to be disposed of in accordance with applicable regulations. Please note all plastic in the containers and measuring cups are high density polyethylene that, when cleaned, are recyclable. Hardened mixed epoxy will not adhere well to polyethylene mixing tray, measuring cups, and plastic tools and may be popped loose and reused after epoxy hardens.

Strategy for Successful Cold Weather Application

1. To successfully apply epoxies at low temperature and have them cure (at low temperature) requires a special strategy. A method must be used to warm the surface to which the coating is to be applied, to warm the air surrounding this surface, and to warm the resins to be applied.

2. Ideally, the application should be made when the surface and air temperature range is 65°–85°F for easy handling as well as for proper and reasonable cure time. For fairly large areas, it is best to "tent" the area, using plastic or other suitable material hung over wood or metal framing. In very cold areas, the wood framing is sometimes insulated with fiberglass batting taped together with duct tape or stapled to a wood frame. Once the area has been tented or sealed off, hot air must be circulated into the area along with proper exhausting of the air, and the surface brought up to reasonable temperature for application. Care should be taken to

engineer the heat source so a fire hazard is not created, and reasonable air circulation is obtained. A surface thermometer should be used to check the surface temperature, as well as checking the surrounding air temperature. If the surface is cold when the epoxy is applied, it will immediately stiffen and become very difficult to spread.

3. Another method of heating smaller areas is to use a bank of one to five electric heat lamps affixed to a 2 × 4, and hung at a distance of about 12"–18" from the surface to which the repair is to be made. These heat lamps may be kept in place during and after the repair to assure positive and reasonably fast cure. These heat lamps may be combined with a heated micro environment around the bonding area to promote the epoxy cure.

4. Epoxy, like motor oil, drops viscosity, or thickness, very quickly when heated. Preheating the material prior to mixing to 80°–90°F will enable easy mixing and faster application. The pot life may be shortened considerably, however, so smaller quantities should be mixed at one time for adequate working time. Keeping a small sample of the mix (one to two ounces) in a warm area will enable one to know when the material is curing satisfactorily.

Before beginning the repair, measure the length of the cracks to be repaired and cut fiberglass tape (with scissors furnished in kit) to fit, overlapping 2" where fiberglass tape is joined.

NOTE: Cut across, following one thread line to avoid unraveling fiberglass. Cut fiberglass each time crack direction is changed so there is at least a 2" overlap on either side of the crack. Try to place no more than 10 linear feet for each mix and keep length of each cut section to 6' or less for easy handling. Roll up each section of fiberglass tape and secure with masking tape furnished with kit; then mark location where each piece will be placed.

Step 1   Surface Preparation

NOTE: It is best to do this preparation the day before the repair to allow concrete to dry thoroughly.

1. Rake out any loose concrete or dirt in cracks. Clean cracks to be repaired with clean water from garden hose with nozzle, using maximum pressure. Broom or squeegee off as much water as possible.

2. While concrete is still damp, put on latex gloves and eye protection. Pour the clean-and-etch material over crack area and use scrub brush furnished to move liquid evenly over area. Scrub any stubborn spots to remove dirt or oil and repeat etching process. Paint brush may be used to apply the clean-and-etch material and neutralizer solution over vertical cracks. Scrub any stubborn dirt while the clean-and-etch material effervesces, 5–8 minutes, then rinse with clean water from garden hose at low pressure.

3. Mix 1 cup (8 ounces in plastic measuring cup furnished) of ammonia with 2½ gallons of water. Measure in the bucket furnished with the kit by filling with water to the bottom ring, then add 8 ounces of ammonia from the bottle furnished in the kit. Pour over cracks to neutralize any remaining acid. Wait 3–5 minutes and rinse again with fresh water. Allow to dry. Sweep, squeegee, or "blow dry" excess water to speed drying. When concrete is dry, clean-and-etched area turns a whitish color. Epoxy repair may now begin.

Step II Preparation of Patching Material

1. Before mixing, place masking tape in a straight line over the concrete approximately 3½" from one side of crack. Then measure 7" from inside edge of tape and place another parallel length of tape. This will furnish a straight edge to the application and provide a gauge of width when spreading the mixed epoxy.

2. Mix only enough epoxy for the length of repair that can be applied in 30 minutes. EXAMPLE: At 70°F, measure in plastic measuring cups provided a total of 1 quart (16 oz. Part A and 16 oz. Part B) for 10–15 linear feet maximum. (Note: One-half gallon covers 30 linear feet 6" wide at 50 mils thickness and does not allow for application loss, tape overlap, etc.) For 5 linear feet, mix one pint (8 oz. Part A and 8 oz. Part B). At 85°F, mix no more than one pint at a time, or use two persons to apply, as product dries faster at high temperatures and slower at lower temperatures. Use the inner plastic tray from the kit for mixing per Figure 2 on kit pail. Wear latex gloves and use wooden mixing paddles furnished, making sure to use separate paddles to stir Part A and Part B to avoid contamination of remaining material in the containers. Use the furnished paint can opener to remove lids from containers. Material must be warm in can. If material has been stored below 70° F, warm to 70°–90° F for easy mixing and application.

STEP III   Mixing Material

Mix equal parts of Part A and Part B in the mixing tray provided until uniform in color, per Figure 3 on kit pail, with another clean wooden paddle furnished in the kit and then continue mixing for the same amount of time (2–3 minutes). **Unmixed product will not dry or will dry soft.**

STEP IV   Applying Material

Quickly apply the mixed epoxy over the crack to be repaired by pouring from the mixing tray over horizontal cracks. Using a 4″ brush, spread to width of slightly less than 7″ between the pieces of masking tape applied in Step II. Then spread evenly with a serrated plastic spreader, using ¼″ side of square spreader as in Step 4. Scrape off excess material, if any, and put back into mixing pail. Coverage varies with topography and porosity of surface. Apply like a heavy house paint.

STEP V   Applying Fiberglass

1. Start to lay 6–12 inches of fiberglass tape. If tape does not wet through from the back side when pressing with square fiberglass spreader, use the smooth side as per Step 5 pictured on kit pail to pull the tape off and apply more epoxy. If too much epoxy is applied, fiberglass tape may move, "wash," or wrinkle easily.

2. Continue laying fiberglass tape onto epoxy resin, overlapping tape end joints 2″. Vertical surfaces are prepared in a similar manner, except that the epoxy must be brushed onto vertical cracks instead of pouring and spreading. Vertical work will be a slower process. **Remove any wrinkles, sags, or other imperfections before material begins to set up. Allow epoxy to set until "tack-free" before proceeding with final application. Clean up plastic tools with soap and water. Discard brush when it hardens and use extra brushes and mixing paddles provided. Epoxy will pop out of mixing tray and measuring cups after hardening and can be re-used.**

STEP VI   Finish Coat

1. After completing placement of tape and allowing to dry (usually overnight), apply a finish coat of epoxy by brush and/or 4″ plastic putty knife applicator to entirely encapsulate the repaired area, making sure all edges of the fiberglass are completely encased in the finish coat. See Figure 6 on kit pail. Before mixing epoxy again

for the finish coat, re-mask at approximately 7" width in the same manner as done originally. **Remove masking tape before material starts to set up or as soon as application is finished.**

2. Allow repairs to dry and cure until "fingernail hard" before allowing traffic. If the area is to be used before surface is hard, cover with 4–6 mil of polyethylene, then cover crack area with plywood. When dry, polyethylene can be removed without harm to application. For clean-up and disposal, see Environmental section of "Instructions for Use" on page 55.

Emergency Response Procedures

Because pesticides are hazardous to human health, safety, and the environment if they inadvertently hit nontarget areas, fires, spills, and other emergency incidents must be responded to quickly to minimize their effect.

The key to reducing the consequences of a spill is to have the proper initial response by the person who first discovers it.

This section outlines the recommended emergency response procedure to follow in case of fire, vehicle accident, spill exposures, or other emergency that involves any bulk pesticides.

General Precautions and First Aid Procedures

General Precautions for All Products

1. Do not take internally.
2. Avoid contact with skin.
3. Do not inhale fumes.
4. In case of contact with eyes, flush with plenty of water and get medical attention.
5. In case of contact with skin, scrub all affected areas with soap and water, remove contaminated clothing, and launder clothing thoroughly before reusing.

First Aid Procedures

1. If poisoning is suspected, contact a physician or the nearest hospital or poison control center.

2. Tell the person contacted of the type and amount of exposure, describe the symptoms, and follow the advice given.

3. See the MSDS for each specific product for more detailed first aid information.

Spill of Pesticides: Control, Containment, and Cleanup

Minor Spill

1. The spilled pesticide should be controlled by stopping the source of the spill. Wear appropriate personal protective equipment.

2. If the spill is outside the containment area and the spilled material starts to spread, it should be contained by diking with sand, soil, or absorbent clay.

3. The contaminated area should be cleared of all personnel except for a small cleanup crew.

4. The cleanup crew should wear appropriate clothing and protective equipment, which may include a respirator, chemical safety goggles, rubber gloves, rubber boots, cotton overalls with a rubber apron, a disposable or rubber suit, and a protective hat.

5. No one should be permitted to enter the area until it is thoroughly decontaminated.

6. If the spill is outside the containment area, the contaminated area should *not* be hosed down immediately, because hosing spreads the spilled pesticide. If the spill is within the containment area, it can be hosed into the sump and transferred to the recycling system.

7. If the spill is outside the containment area, activated charcoal, absorptive clay, vermiculite, pet litter, sweeping compound, or lime should be thrown over the entire spill. Enough absorbent material should be used to soak up as much of the liquid as possible. If possible, leave the absorbent in contact with the liquid for at least one hour.

8. All the nonrecyclable absorbed material should be swept or shoveled into a large leakproof open-head recovery drum. Repeat the cleanup process one or two more times until all visible residue of spilled material is removed.

9. After cleanup, the area should be washed with soap and water.

10. If the spill is outside the containment area, absorb the wash solution with absorbent and sweep up. If the spill is inside the containment area, hose it into a sump and recycle.

11. Finally, the area should be rinsed with water to wash away any remaining material. The rinse water should be collected and held for disposal or recycling.

12. Discard contaminated clothing (that cannot be cleaned), brooms, and so forth. This is especially important when cleaning up the more toxic pesticides.

13. Once all nonrecyclable material is cleaned up and placed in the recovery drum, seal the drum and arrange for disposal in accordance with state and local regulations.

14. Avoid contamination when removing protective clothing and equipment.

    Note: If a spill occurs in a pesticide storage area, the area should be checked carefully to see if any other pesticides have been contaminated by the spill. If so, these pesticides should also be disposed of or arrangements should be made for their return to the manufacturer (if the product can be salvaged). Any disposal should be carried out in accordance with local, state, and federal regulations.

Major Spill

1. If a cleanup job appears to be too big to handle, or if the correct procedure is unclear, telephone Chem-Trec at (800) 424-9300.

2. If a spill occurs on a highway, call the state highway patrol or the local sheriff's office for assistance. Do not leave the area until responsible assistance arrives.

3. Do not walk in spilled pesticide. Prevent vehicles from driving over spilled material.

4. Do not handle leaking containers or enter a tank or van without appropriate protective clothing and a respirator or self-contained breathing apparatus (SCBA).

5. Do not allow anyone to smoke near the spilled pesticide or provide any other source of ignition.

6. Dike the spill to prevent runoff of pesticide into any nearby waterways, ditches, streams, ponds, and so on.

7. Wear protective clothing and equipment appropriate for the pesticide spilled.

8. Follow the control, containment, and cleanup procedures outlined under "Minor Spill."

9. If soil is contaminated by the spill, the contaminated soil should be removed to a depth of two inches below the level of contamination.

10. Do not hose down the spill until all spill material is cleaned up and the area has been decontaminated.

Fires Involving Pesticides

General Rules to Follow in Case of Fire

1. Call the fire department and clear all personnel from the building to a safe distance *upwind* from smoke and fumes. Isolate the area if necessary.

2. Protect the firefighters. Fire-fighting personnel should wear impervious clothing, including liquid-proof hats, coats, trousers, and full-face, self-contained breathing apparatus, rubber boots, and rubber gloves to prevent contact with the pesticide.

3. A self-contained breathing apparatus should be used while fighting a pesticide fire to provide respiratory protection against both toxic vapors and an oxygen-deficient atmosphere (oxygen less than 19.5 percent).

   Caution: Do not use gas masks with canisters while fighting the fire. Protection against oxygen deficiency is *not* provided by these respirators or by gas masks ordinarily used to protect against pesticide inhalation.

4. Use standard organic chemical fire-fighting techniques to extinguish fires involving pesticides. Use dry chemicals, foam, or carbon dioxide.

5. Confine and isolate the fire and contamination. Keep stored pesticide containers cool, if possible, with water spray. Use water spray to diminish smoke and vapor if necessary. Trench or dike around the area, if possible, to contain contaminated water.

   Note: Use only as much water as necessary, because excess water aggravates the cleanup problem. If the structure cannot be saved, it may be better to let it burn, because additional water application could result in extensive contaminated water runoff or incomplete combustion of chemicals resulting in a release of toxic compounds into the air. This decision needs to be made by the commanding fire official at the scene.

6. Evacuate persons who might be endangered by the fire or the resulting smoke and fumes. *Keep away from smoke from burning pesticides.*

Cleanup Procedures

1. Cleanup procedures are dictated by the type of material involved (e.g., solid material such as wood or steel, vs. pesticide containers and residue).

2. Cleanup procedures are also determined by the amount of debris and water for disposal. If large quantities of water or pesticides are involved, contact Chem-Trec for instructions on deactivation and/or disposal.

3. Follow the cleanup procedures outlined under "Minor Spill."

Disposal

1. Damaged or unsalvageable product, wash water, spill residue, and debris must be disposed of by methods consistent with local, state, and federal health and environmental regulations.

2. Unsalvageable pesticides must be disposed of in an approved chemically secure landfill or destroyed by incineration or chemical degradation.

3. Any contaminated soil, building materials, or debris must also be disposed of or destroyed in the same manner as outlined above.

APPENDIX II

Block Diagrams

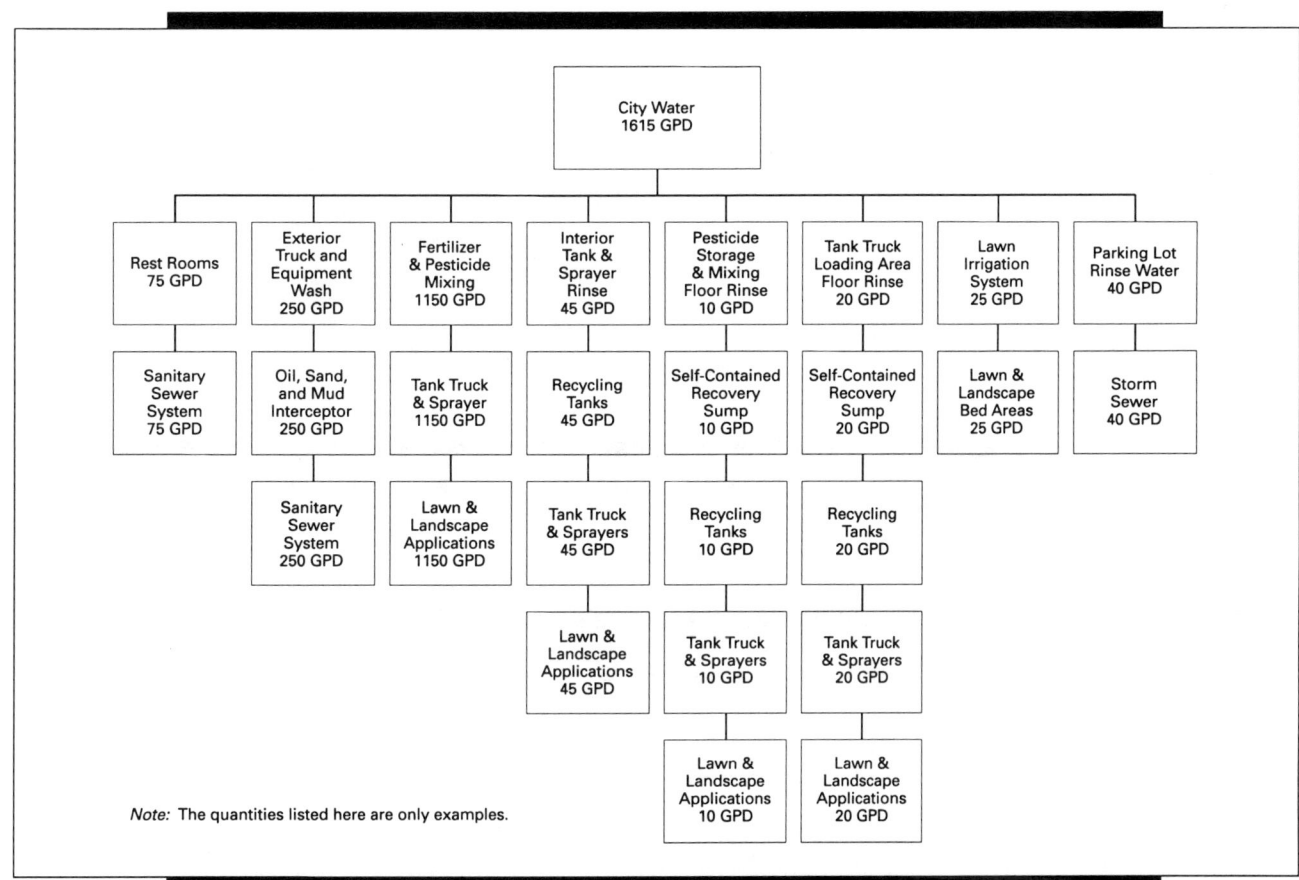

Note: The quantities listed here are only examples.

Water Usage, Wastewater Generation, and Rinse-Water Recycling—Active-Season Usage, March Through October

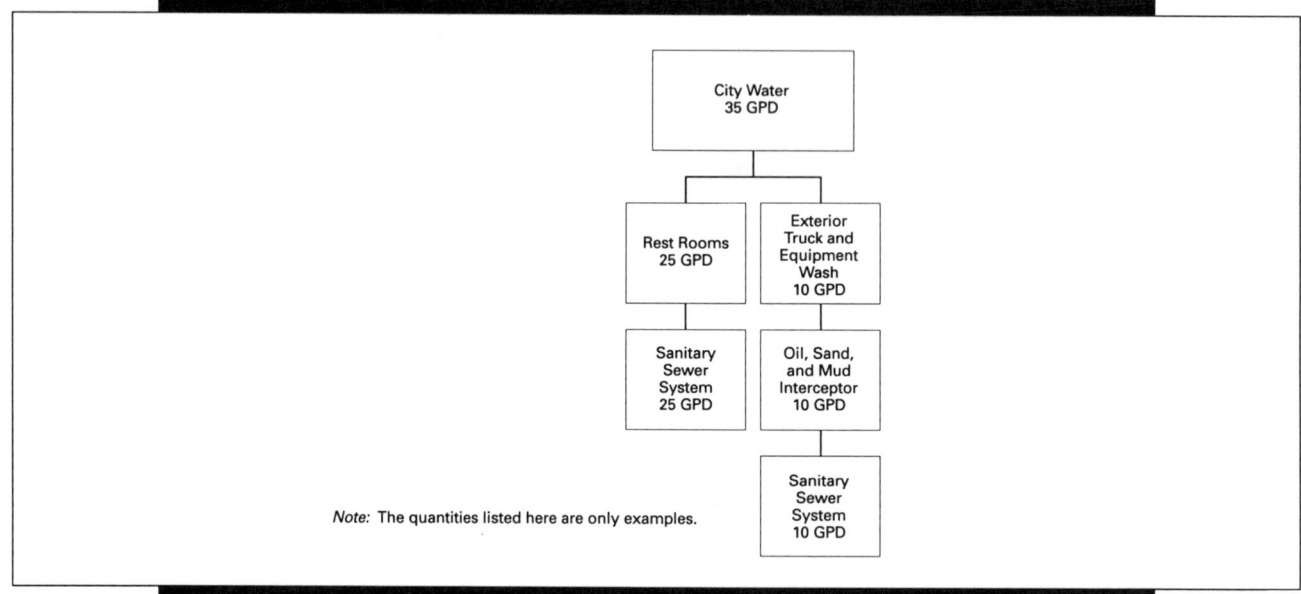

City Water
35 GPD
Rest Rooms
25 GPD
Exterior
Truck and
Equipment
Wash
10 GPD
Sanitary
Sewer
System
25 GPD
Oil, Sand,
and Mud
Interceptor
10 GPD
Sanitary
Sewer
System
10 GPD
Note: The quantities listed here are only examples.

Water Usage, Wastewater Generation, and Rinse-Water Recycling—Off-Season Usage, November Through February

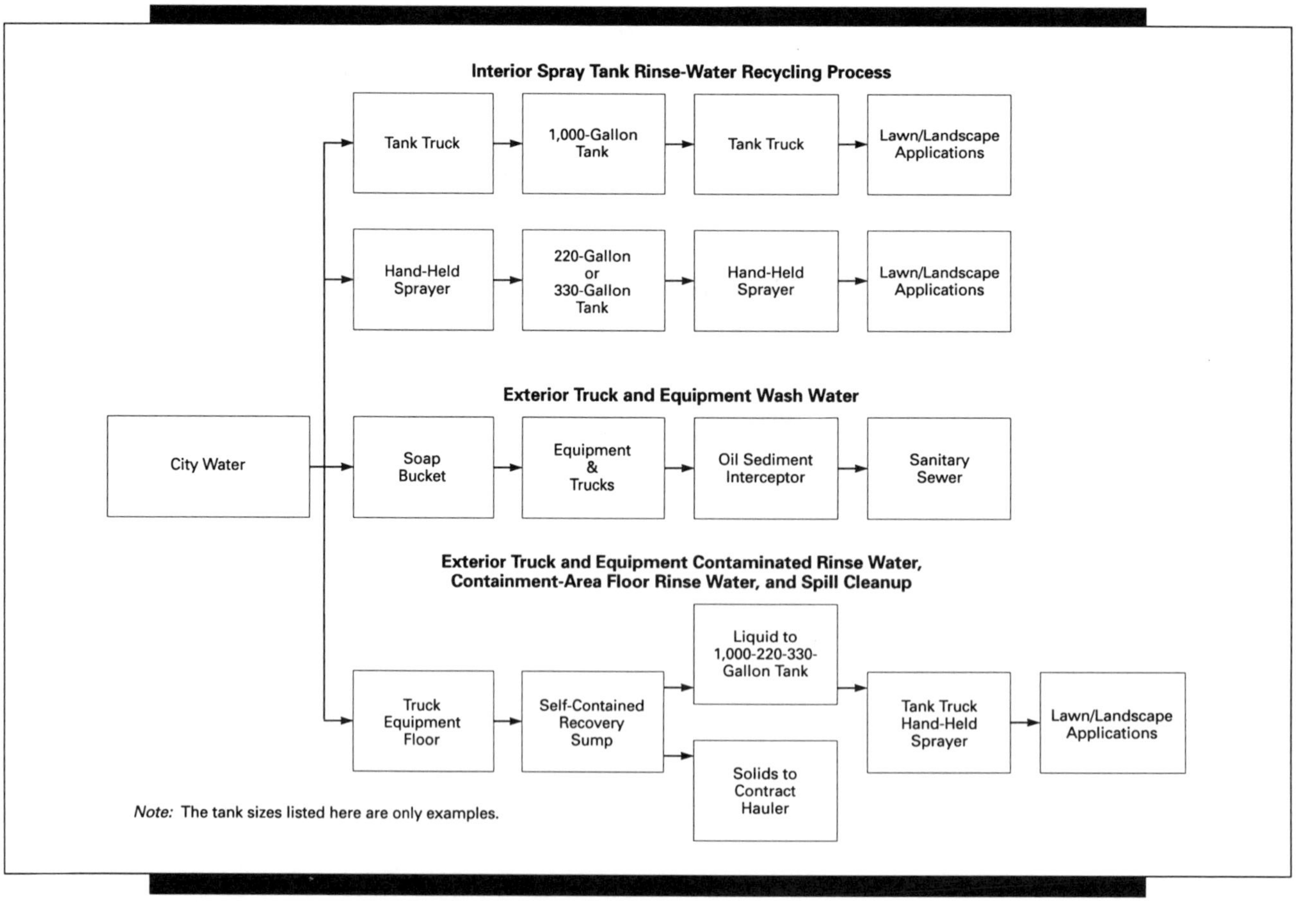

Interior Spray Tank Rinse-Water Recycling Process
Tank Truck
1,000-Gallon
Tank
Tank Truck
Lawn/Landscape
Applications
Hand-Held
Sprayer
220-Gallon
or
330-Gallon
Tank
Hand-Held
Sprayer
Lawn/Landscape
Applications
Exterior Truck and Equipment Wash Water
City Water
Soap
Bucket
Equipment
&
Trucks
Oil Sediment
Interceptor
Sanitary
Sewer
Exterior Truck and Equipment Contaminated Rinse Water,
Containment-Area Floor Rinse Water, and Spill Cleanup
Liquid to
1,000-220-330-
Gallon Tank
Truck
Equipment
Floor
Self-Contained
Recovery
Sump
Tank Truck
Hand-Held
Sprayer
Lawn/Landscape
Applications
Solids to
Contract
Hauler
Note: The tank sizes listed here are only examples.

Interior Spray Tank Rinse-Water Recycling Process

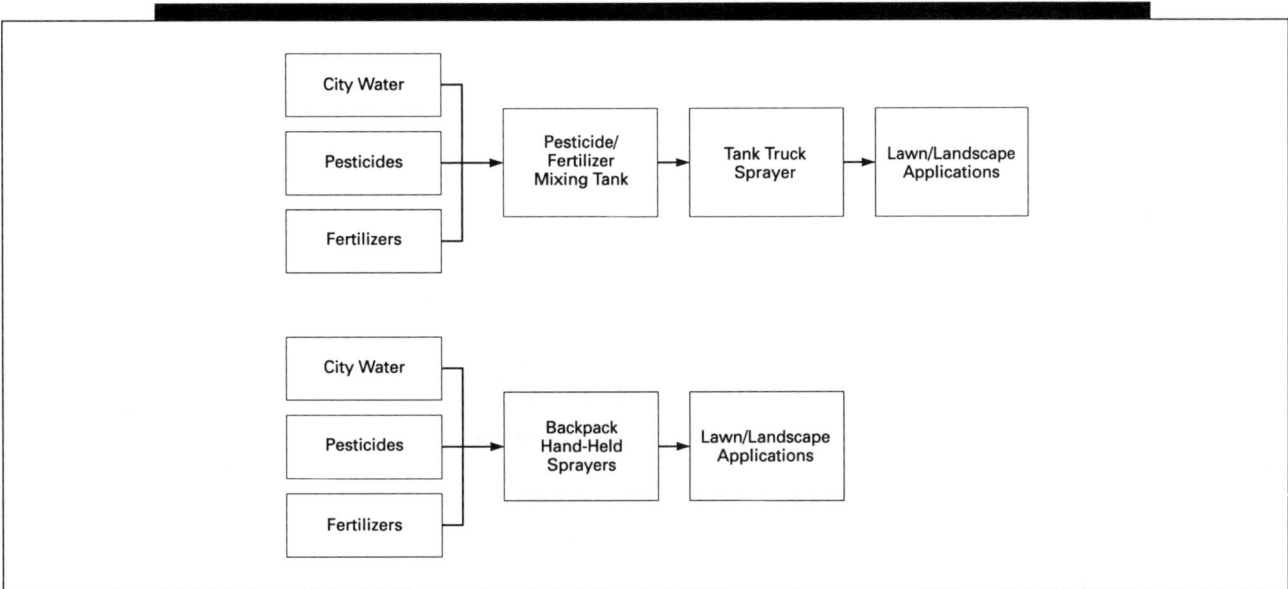

Process of Finished Product Formulation and Usage

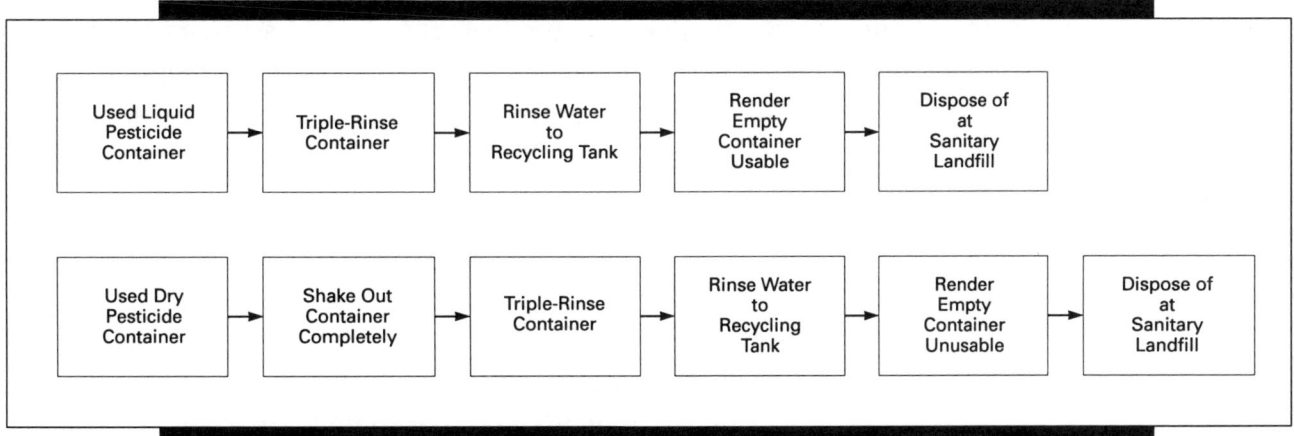

Pesticide Empty Container Disposal Procedure

Design Calculations, Formulas, and Specifications

Procedure for Calculating the Volume of a Tank

For a Flat-Bottom Tank:

A. Volume of a cylinder in cubic feet (ft.^{3}) = radius squared (r^2) × 3.1416 × height (h) or diameter squared (d^2) × 0.7854 × height (h)

   Note: (1) If you know the circumference, divide by 3.1416 to get the diameter (d).

   (2) If you take an outside measurement, subtract 1″ from diameter (d) to adjust for the thickness of the wall of the tank.

B. Gallons = cubic feet (ft.^{3}) × 7.48.

   Note: This value is based on the fact that one cubic foot of water or one cubic foot of any liquid at room temperature equals 7.48 gallons. They will just have different weights.

For a Cone-Bottom Tank:

A. Volume of a cylinder (ft.^{3}) + volume of a cone (ft.^{3}) = volume of the tank:

$$(r^2 \times 3.1416 \times h) + (r^2 \times 1.0472)$$

or

$$(d^2 \times 0.7854 \times h) + (r^2 \times 3.1416 \times h/3)$$

B. Gallons = (ft.^{3} of cylinder × 7.48) + (ft.^{3} of cone × 7.48).

Procedure for Calculating the Amount of a Product in Water

1. Calculate total acre-feet (surface acres × ave. depth):

   Example: A pond is 756' long, 169' wide and has an average depth of 2.66'.

   a. surface acres = × 756' × 169' ÷ 43560 ft.^{2}
      = 2.93

   b. surface acres × ave. depth = acre-feet
      2.93 × 2.66' = acre-feet
      7.79 = acre-feet

2. 2.72 lbs. of any material in one acre-foot of water = 1 ppm.

3. Calculate the total pounds of active ingredient of each product in the pond (ppm × 2.72 × acre-feet):

   Example: 2 ppm of an herbicide is detected in the water of the pond listed above.

   Answer: 2 ppm × 2.72 × 7.79 acre-feet = 42.38 lbs. of active ingredients in the pond.

Water Concentration Conversions

1. ppm in water × 2.72 × acre-feet = lbs. of material in water.

2. One ft.^{3} of water = 62.5 lbs. or 7.48 gals.

3. One gal. of water = 8.34 lbs.

4. Acre-feet = surface acres × ave. water depth.

5. 1 oz./1000 ft.^{2} = 2.5 lbs./acre.

6. 2.72 lbs. of anything = 1 ppm/acre-foot.

7. ug/L = ppb

8. $\dfrac{\text{ug/L}}{1000}$ = ppm

9. mg/L = ppm

10. mg/L × 1000 = ppb = ug/L

11. $\dfrac{\text{lbs a.i.}}{\text{lbs. of diluent}}$ × 1,000,000 = ppm of any product in a spray mixture

12. lbs. a.i. of a product in one gallon of a mixture = actual wt./gal. of the product × % a.i. present.

    Note: If the actual weight per gallon is unknown, it can be calculated by multiplying the specific gravity of the product sample × 8.34.

Soil Volume Conversions

1. $\dfrac{\text{ft.}^3}{27} = \text{yds.}$

2. $90 \times \text{ft.}^3 = \text{lbs. of soil. } (90 \text{ lbs.}/\text{ft.}^3)$

3. Rule of thumb—it costs \$125/T to haul soil to a landfill.

4. One large dump truck (semi-trailer) can usually carry 17 yards of soil.

5. To convert ppm to lbs. in a given volume of soil:

   a. $\dfrac{\text{ppm}}{1,000,000} = \dfrac{\text{lbs.}}{90 \times \text{ft.}^3}$

   b. lbs. of active ingredient $= \dfrac{\text{ppm} \times (90 \times \text{ft.}^3)}{1,000,000}$

Tank Capacities

Table A is useful in estimating the contents of partially filled horizontal cylindrical tanks with flat ends.

For computing values not found in the table, use the procedure of the following example.

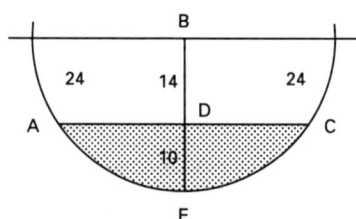

Dimensions used in calculation of capacities of horizontal cylindrical tanks

Assume tank diameter equals 48 in., filled with liquid to a depth of 10 in. Find the volume in U.S. gallons per foot of tank length.

Area ACE = area ABCE − area ABC

Area ABCE $= \left(\dfrac{2\angle \text{ABD}}{360} \right) \times$ area of circle, and $\angle$ABD is found from its cosine, which is $^{14}\!/_{24}$.

$\therefore \angle \text{ABD} = 54.25°$.

$$\text{Area ABC} = 14 \times 24 \times \sin \angle \text{ABD}$$

$$\text{or } 14 \times 24 \times \sin 54.25 = 14 \times 24 \times .8116 = 272.7$$

$$\text{Area ABCE} = \frac{108.50}{300} \times -x(24)^2 = 545.4$$

$$\therefore \quad \text{Area ACE} = 272.7$$

$$\therefore \quad \text{Vol. (U.S. gal.) per ft. of length} = \frac{272.7 \times 12}{231} = 14.17.$$

TABLE A. Capacities of Horizontal Cylindrical Tanks

Contents given in U.S. gallons for 1 ft. of length
Flat ends

Diameter of tank, in.	Depth of liquid, in.																
	3	6	9	12	15	18	21	24	27	30	33	36	39	42	45	48	51
12	1.15	2.94	4.73	5.88													
18	1.45	3.86	6.61	9.36	11.77	13.22											
24	1.70	4.60	8.05	11.75	15.45	18.90	21.80	23.50									
30	1.91	5.23	9.27	13.72	18.36	23.00	27.45	31.49	34.81	36.72							
36	2.12	5.79	10.34	15.43	20.85	26.44	32.03	37.45	42.54	47.09	50.76	52.88					
42	2.28	6.31	11.31	16.97	23.07	29.46	35.99	42.51	48.90	55.00	60.66	65.66	69.69	71.97			
48	2.45	6.78	12.20	18.38	25.10	32.20	39.54	47.01	54.47	61.81	65.91	75.63	81.81	87.23	91.56	94.01	
54	2.60	7.22	13.04	19.68	26.97	34.72	42.80	51.08	59.49	67.90	76.18	84.26	92.01	99.30	105.94	111.76	116.35
60	2.75	7.64	13.82	20.91	28.72	37.06	45.82	54.83	64.11	73.45	82.78	92.01	101.07	109.83	118.17	125.93	133.07
66	2.89	8.04	14.56	22.07	30.37	39.28	48.65	58.39	68.42	78.59	88.87	99.14	109.31	119.34	129.05	135.45	147.36
72	3.02	8.42	15.26	23.17	31.92	41.36	51.32	61.71	72.45	83.41	94.54	105.76	116.93	128.11	139.07	149.81	160.20
78	3.15	8.73	15.94	24.21	33.41	43.34	53.86	64.87	76.27	87.97	99.90	111.97	124.12	136.27	143.34	160.27	171.97
84	3.26	9.12	16.57	25.24	34.85	45.24	56.29	67.87	79.91	92.30	104.98	117.85	130.87	143.95	157.03	170.05	182.92
90	3.43	9.46	17.20	26.20	36.21	47.05	58.61	70.75	83.39	96.43	109.81	123.45	137.23	151.22	165.25	179.27	193.21
96	3.50	9.79	17.80	27.13	37.52	48.81	60.84	73.52	86.73	100.39	114.44	128.79	143.40	158.17	173.07	183.01	202.95
102	3.61	10.10	18.37	28.01	39.00	50.49	62.99	76.13	89.94	104.20	118.89	133.92	149.23	164.81	180.53	196.36	212.25
108	3.71	10.39	18.94	28.90	40.03	52.14	65.09	78.74	93.04	107.87	123.17	138.87	154.89	171.19	187.71	204.37	221.15

Diameter of tank, in.	Depth of liquid, in.																
	54	57	60	63	66	69	72	75	78	81	84	87	90	93	96	102	108
54	118.38																
60	139.25	144.14	146.89														
66	155.66	163.17	169.69	174.84	177.73												
72	170.16	179.60	188.35	196.26	203.10	208.50	211.52										
78	183.37	194.38	204.90	214.83	224.03	232.30	239.46	245.09	248.24								
84	195.60	207.99	220.03	231.61	242.66	253.05	262.66	271.33	278.78	284.64	287.90						
90	207.03	220.65	234.06	247.10	259.74	271.83	283.44	294.28	304.29	313.29	321.03	327.06	330.49				
96	217.85	232.62	247.23	261.58	275.63	289.29	302.50	315.18	327.21	338.50	348.89	358.22	366.23	372.52	376.02		
102	228.12	243.95	259.67	275.23	290.56	305.59	320.28	334.54	348.30	361.49	373.99	385.48	396.47	406.11	414.33	424.48	
108	238.05	254.75	271.53	285.19	304.71	321.01	337.03	352.73	368.03	382.86	397.16	410.81	423.76	435.87	447.00	465.51	475.99

Tank Capacities, Horizontal Cylindrical

TABLE B. Contents of Tanks with Flat Ends When Filled to Various Depths

Contents in U.S. gallons per 1 ft of length.

Diameter of tank, in.	Full tank	3"	6"	9"	12"	15"	18"	21"	24"	27"	30"	33"	36"	39"	42"	45"	48"	51"	54"	57"	60"
12"	5.89	1.15	2.94	...	...																
18"	13.22	1.45	3.86	6.61	...																
24"	23.50	1.70	4.60	8.05	11.75																
30"	36.72	1.91	5.23	9.27	13.72	18.36	...	...													
36"	52.88	2.12	5.79	10.34	15.43	20.85	26.44	...													
42"	71.97	2.28	6.31	11.31	16.97	23.07	29.47	35.99													
49"	94.01	2.45	6.76	12.20	18.38	25.10	32.20	39.54	47.00	...	...										
54"	118.98	2.60	7.22	13.04	19.68	26.97	34.72	42.80	51.08	59.49	...										
60"	146.89	2.75	7.64	13.82	20.91	28.72	37.06	45.82	54.87	64.11	73.44										
66"	177.73	2.89	8.04	14.56	22.07	30.37	39.28	48.65	58.39	68.41	78.59	88.86	...	...							
72"	211.52	3.02	8.42	15.26	23.17	31.92	41.36	51.32	61.71	72.45	83.41	94.54	105.76	...							
78"	248.24	3.15	8.78	15.94	24.21	33.41	43.34	53.86	64.87	76.27	87.97	99.90	111.97	124.13							
84"	287.90	3.26	9.12	16.57	25.24	34.85	45.24	56.29	67.87	79.91	92.30	104.99	117.85	130.87	143.95	...	...				
90"	330.49	3.43	9.46	17.20	26.20	36.21	47.05	58.61	70.75	83.39	96.43	109.81	123.45	137.28	151.23	165.25	...				
96"	376.02	3.50	9.79	17.80	27.13	37.52	48.81	60.84	73.52	86.73	100.39	114.44	128.79	143.40	158.17	173.06	183.01				
102"	424.50	3.61	10.10	18.37	28.01	39.00	50.49	62.99	76.18	89.94	104.20	118.89	133.92	149.25	164.81	180.53	196.37	212.25	...	...	...
108"	476.10	3.71	10.39	18.94	28.90	40.03	52.14	65.09	78.74	93.04	107.87	123.17	138.87	154.89	171.19	187.71	204.37	221.14	238.05	...	...
114"	530.25	3.78	10.74	19.49	29.75	41.22	53.73	67.10	81.24	96.05	111.43	127.31	143.63	160.33	177.33	194.60	212.05	229.65	247.37	265.13	...
120"	587.54	3.91	10.98	20.02	30.57	42.39	55.26	69.06	83.65	98.95	114.87	131.32	149.25	165.58	183.27	201.24	219.46	237.87	256.43	275.08	293.77

To ascertain the contents of a tank over one-half full: Let h = depth of unfilled portion. Find from the table the quantity corresponding to a depth "h." Subtract this quantity from the contents of a full tank.

TABLE C. Contents of Standard Dished Heads When Filled to Various Depths

Contents in U.S. gallons for *one head only*. This table is only approximate but close enough for practical use.
As prepared by Lukens Steel Company.

Diameter of tank, in.	Full tank	3"	6"	9"	12"	15"	18"	21"	24"	27"	30"	33"	36"	39"	42"	45"	48"	51"	54"	57"	60"
12"	0.40	0.05	0.20	...	...																
18"	1.36	0.07	0.32	0.63	...																
24"	3.22	0.08	0.41	0.95	1.61																
30"	6.30	0.10	0.49	1.18	2.10	3.15	...	...													
36"	10.88	0.11	0.55	1.39	2.54	3.92	5.44	...													
42"	17.28	0.12	0.68	1.59	2.94	4.64	6.57	8.64													
43"	25.79	0.13	0.68	1.75	3.31	5.29	7.62	10.19	12.89	...	...										
54"	36.72	0.14	0.74	1.90	3.64	5.91	8.60	11.65	14.95	18.36	...										
60"	50.37	0.14	0.82	2.07	2.98	6.49	9.54	13.03	16.87	20.96	25.18										
66"	67.04	0.15	0.83	2.19	4.25	6.96	10.35	14.30	18.68	23.43	28.42	33.52	...	...							
72"	87.04	0.16	0.88	2.32	4.52	7.47	11.15	15.49	20.38	25.74	31.46	37.43	43.52	...							
78"	110.66	0.17	0.93	2.44	4.79	7.97	11.94	16.65	22.02	27.97	34.39	41.16	48.20	55.33							
84"	138.22	0.18	0.98	2.59	5.07	8.44	12.69	17.76	23.60	30.11	37.19	44.75	52.67	60.83	69.11	...	...				
90"	170.01	0.18	1.00	2.68	5.33	8.91	13.44	18.86	25.12	32.18	39.90	48.22	56.99	66.10	75.52	85.00	...				
96"	206.32	0.20	1.07	2.83	5.59	9.36	14.14	19.90	26.60	34.17	45.52	51.53	61.13	71.22	81.65	92.34	103.16				
102"	247.48	0.22	1.14	3.01	5.89	9.87	14.92	21.01	28.11	36.18	45.19	54.91	65.31	76.29	87.73	99.56	111.59	123.74	...	...	...
108"	293.77	0.20	1.13	3.03	6.04	10.21	15.50	21.93	29.47	38.03	47.56	57.97	69.14	81.05	93.53	106.47	119.76	133.26	146.88	...	...
114"	345.51	0.21	1.16	3.12	6.25	10.55	16.06	22.80	30.70	39.73	49.81	60.80	72.65	85.61	99.00	113.07	127.56	142.41	157.51	172.75	...
120"	402.27	0.21	1.19	3.23	6.47	10.93	16.69	23.70	31.96	41.43	52.04	63.73	76.40	89.95	104.32	119.39	135.04	151.15	167.62	184.32	201.13

To ascertain the contents of a head over one-half full: Let h = depth of unfilled portion. Find from the table the quantity corresponding to a depth "h." Subtract this quantity from the contents of a full head.

Quantities stored in tanks can be readily determined from the dimensions of the tank and the depth of the liquid in the tank. Capacities of rectangular tanks are given in Table D and of vertical tanks in Table E. The contents of horizontal cylindrical tanks will be influenced by the shape of the tank ends, which may be convex, straight, or concave. For rough estimating, the effect of the ends may be neglected. When a horizontal cylindrical tank is only partly filled, a deduction must be made from the total contents of the tank equal to the unfilled portion above the liquid level.

If the tank is filled to a point above the axis, then subtract from the total contents an amount equal to the contents of a space having the same length as the tank and an end area equal to the unwetted segment of the tank end. If the tank is filled to a point below the axis, then the contents become equal to that obtained by multiplying the length of the tank by the area of the wetted segment of the end of the tank.

TABLE D. Capacities (U.S. Gallons) of Rectangular Tanks for Each Foot of Liquid
231 cu. in. = 1 U.S. gal.;* 1 cu. ft. = 1728 cu. in. = 7.4805 U.S. gal.

Tank width, in.	6″ / 0.5′	12″ / 1′	18″ / 1′-6″	24″ / 2′	30″ / 2′-6″	36″ / 3′	42″ / 3′-6″	48″ / 4′	54″ / 4′-6″	60″ / 5′	66″ / 5′-6″	72″ / 6′	78″ / 6′-6″	84″ / 7′
6	1.87	3.74	5.61	7.48	9.35	11.22	13.09	14.96	16.83	18.70	20.57	22.44	24.31	26.18
12	3.74	7.48	11.22	14.96	18.70	22.44	26.18	29.92	33.66	37.40	41.14	44.88	48.62	52.36
18	5.61	11.22	16.83	22.44	28.05	33.66	39.27	44.88	50.49	56.10	61.71	67.32	72.93	78.55
24	7.48	14.96	22.44	29.92	37.40	44.88	52.36	59.84	67.32	74.81	82.29	89.77	97.25	104.73
30	9.35	18.70	28.05	37.40	46.73	56.10	65.45	74.81	84.16	95.51	102.86	112.21	121.56	130.91
36	11.22	22.44	33.66	44.88	56.10	67.32	78.55	89.77	100.99	112.21	123.43	134.65	145.87	157.09
42	13.09	26.18	39.27	52.36	65.45	78.55	91.64	104.73	117.82	130.91	144.00	157.09	170.18	183.27
48	14.96	29.92	44.88	59.84	74.81	89.77	104.73	119.69	134.65	149.61	164.57	179.53	194.49	209.45
54	16.83	33.66	50.49	67.32	84.16	100.99	117.82	134.65	151.48	168.31	185.14	201.97	218.81	235.64
60	18.70	37.40	56.10	74.81	93.51	112.21	130.91	149.61	168.31	187.01	205.71	224.42	243.12	261.82
66	20.57	41.14	61.71	82.29	102.86	123.43	144.00	164.57	185.14	205.71	226.29	246.86	267.43	288.00
72	22.44	44.88	67.32	89.77	112.21	134.65	157.09	179.53	201.97	224.42	246.86	269.30	291.74	314.18
78	24.31	48.62	72.94	97.25	121.56	145.87	170.18	194.49	218.81	243.12	267.43	291.74	316.05	340.36
84	26.18	52.36	78,55	104.73	130.91	157.09	183.27	209.45	235.64	261.82	238.00	314.18	340.36	366.55
90	28.05	56.10	84.16	112.21	140.26	168.31	196.36	224.42	252.47	280.52	308.58	336.62	364.68	392.73
96	29.92	59.84	89.77	119.69	149.61	179.53	209.45	239.38	269.30	299.22	329.14	359.06	388.99	418.91

Tank width, in.	90″ / 7′-6″	96″ / 8′	102″ / 8′-6″	108″ / 9′	114″ / 9′-6″	120″ / 10′	126″ / 10′-6″	132″ / 11′	138″ / 11′-6″	144″ / 12′	150″ / 12′-6″	156″ / 13′	162″ / 13′-6″	168″ / 14′
6	28.05	29.92	51.79	33.66	35.53	37.40	39.27	41.14	43.01	44.88	46.75	48.62	50.49	52.36
12	56.10	59.84	63.58	67.32	71.06	74.81	78.55	82.29	86.03	89.77	93.51	97.25	100.99	104.73
18	84.16	89.77	95.38	100.99	106.60	112.21	117.82	123,43	129.04	134.65	140.26	145.87	151.48	157.09
24	112.21	119.69	127.17	134.65	142.13	149.61	157.09	164.57	172.05	179.53	187.01	194.49	201.97	209.45
30	140.26	149.61	158.96	168.31	177.66	187.01	196.36	205.71	215.06	224.42	233.77	243.12	252.47	261.82
36	168.31	179.53	190.75	201.97	213.19	224.42	235.64	246.86	258.08	269.30	280.52	291.74	302.96	314.18
42	196.36	209.45	222.55	235.64	248.73	261.82	274.91	288.00	301.09	314.18	327.27	340.36	353.45	366.55
48	224.42	239.38	254.34	269.30	284.26	299.22	314.18	329.14	344.10	359.07	374.03	388.99	403.95	418.91
54	252.47	269.30	286.13	302.96	319.79	336.62	353.45	370.29	387.12	403.95	420.78	437.61	454.44	471.27
60	280.52	299.22	317.92	336.62	355.32	374.03	392.73	411.43	430.13	448.83	467.53	486.23	504.94	523.64
66	308.57	329.14	349.71	370.29	390.86	411.43	432.00	452.57	473.14	493.71	514.29	534.86	555.43	576.00
72	336.62	359.06	381.51	403.95	426.39	448.83	471.27	493.71	516.16	538.60	561.04	583.48	605.92	628.36
78	364.68	388.99	413.30	437.61	461.92	486.23	510.55	534.86	559.17	583.48	607.79	632.10	656.42	680.73
84	392.73	418.91	445.09	471.27	497.45	523.64	549.82	576.00	602.18	628.36	654.55	680.73	706.91	733.09
90	420.78	448.83	476.88	504.94	532.99	561.04	589.09	617.14	645.19	673.25	701.30	729.35	757.40	785.45
96	448.83	478.75	508.68	538.60	568.52	598.44	628.36	658.29	688.21	718.13	748.05	777.97	807.90	837.82

*For Imperial gallons, multiply above capacities by 1.2.

TABLE E. Capacities of Vertical Cylindrical Tanks in U.S. Gallons								
Diameter		Gal/ft. depth	Diameter		Gal/ft. depth	Diameter		Gal/ft. depth
Ft.	In.		Ft.	In.		Ft.	In.	
0	0		10	0	587.52	20	0	2350.1
	3	0.37		3	617.26		3	2409.2
	6	1.47		6	647.74		6	2469.1
	9	3.31		9	678.95		9	2529.6
1	0	5.88	11	0	710.90	21	0	2591.0
	3	9.18		3	743.58		3	2653.0
	6	13.22		6	776.99		6	2715.8
	9	17.99		9	811.14		9	2779.3
2	0	23.50	12	0	846.03	22	0	2843.6
	3	29.74		3	881.64		3	2905.6
	6	36.72		6	918.00		6	2974.3
	9	44.43		9	955.08		0	3040.8
3	0	52.88	13	0	992.91	23	0	3108.0
	3	62.06		3	1031.5		3	3175.9
	6	71.97		6	1070.8		6	3244.6
	9	82.62		9	1110.8		9	3314.0
4	0	94.00	14	0	1151.5	24	0	3384.1
	3	106.12		3	1193.0		3	3455.0
	6	118.97		6	1235.3		6	3526.6
	9	132.56		9	1278.2		9	3598.9
5	0	146.88	15	0	1321.9	25	0	3672.0
	3	161.93		3	1366.3		3	3745.8
	6	177.72		6	1411.5		6	3820.3
	9	194.25		9	1457.4		9	3895.6
6	0	211.51	16	0	1504.1	26	0	3971.6
	3	229.50		3	1551.4		3	4048.4
	6	248.23		6	1599.5		6	4125.9
	9	267.69		9	1648.4		9	4204.1
7	0	287.88	17	0	1697.9	27	0	4283.0
	3	308.81		3	1748.2		3	4362.7
	6	330.48		6	1799.3		6	4443.1
	9	352.88		9	1851.1		9	4524.3
8	0	376.01	18	0	1903.6	28	0	4606.1
	3	399.88		3	1956.8		3	4688.8
	6	424.48		6	2810.8		6	4772.1
	9	449.82		9	2065.5		9	4856.2
9	0	475.89	19	0	2120.9	29	0	4941.0
	3	502.70		3	2177.1		3	5026.6
	6	530.24		6	2234.0		6	5112.9
	9	553.51		9	2291.7		9	5199.9

Concrete Dike Calculations

Concrete is measured and sold by cubic yard (27 cubic feet). To calculate the amount of concrete required for the dike, figure the volume of the forms in cubic feet (thickness × width × length) and divide by 27 to obtain the total number of cubic yards.

For example, to pour a 4-inch thick concrete pad that is 20 feet wide and 40 feet long, calculate the cubic footage by multiplying the thickness, 4 inches (or $\frac{1}{3}$ of a foot) times 20 feet (width) times 40 feet (length). This multiplies out to $266\frac{2}{3}$, or 267 cubic feet. Converting this to cubic yards, $267/27 = 9\frac{24}{27}$, or 10 cubic yards of concrete to complete the driveway.

The chart below can be used in estimating amounts of concrete. For the $4'' \times 20' \times 40'$ concrete pad, read down the left column of the table to the depth of the concrete in inches, which is 4 inches. Read across to find that one cubic yard at that depth will fill 81 square feet. The area of the pad is 800 square feet (length × width). Next divide the total square feet by the number of square feet covered by one cubic yard of cement:

$$800/81 = 9.88$$

Roughly 10 cubic yards of concrete will be needed for the pad.

Calculation

This table indicates the area in square feet that one cubic yard of concrete will fill for a variety of thicknesses. For example, one cubic yard of concrete will fill a form area of 54 square feet for an area that is 6 inches thick.

Concrete Estimating 1 Cubic Yard of Concrete Will Fill

Thickness (Inches)	Sq. Ft.	Thickness (Inches)	Sq. Ft.	Thickness (Inches)	Sq. Ft.
1	324	5	65	9	36
1½	216	5½	59	9½	34
2	162	6	54	10	32.5
2½	130	6½	50	10½	31
3	108	7	46	11	29.5
3½	93	7½	43	11½	28
4	81	8	40	12	28
4½	72	8½	38		

Labor

Labor is estimated to be 2–2.3 man-hours per cubic yard of concrete. Therefore, to construct the pad $20' \times 40' \times 4''$ thick, it would take 20 man-hours for construction of forms and pouring cement.

Materials Estimate

Cost of Concrete per Cubic Yard (4,000 lbs./yd.^{3})
- 1–3 cu. yds. = $50–$60
- 3–10 cu. yds. = $45–$55
- 10–20 cu. yds. = $42–$50

Design Calculations for Multiple Tank Storage

Tank Data

    1. Tank #1, capacity: C1 = _____ gal. (largest volume)

                diameter: D1 = _____ ft.

    2. Tank #2, capacity: C2 = _____ gal.

                diameter: D2 = _____ ft.

    3. Tank #3, capacity: C3 = _____ gal.

                diameter: D3 = _____ ft.

Sketch of Containment Area with Tanks Shown (Sketch Available Area to Place Tanks)

-ft.

-ft.

Calculate Available Surface Area within the Containment Wall Area:

(AC = containment area)

AC = _____ ft. × _____ ft. = _____ ft.^{2} inside wall area

Calculate Volume of Largest Tank Plus 10%: (Lg = gallons of largest tank)
(Vt = volume of largest tank)

Lg = C1 = _____ gal. + _____ (10%) gal. = _____ gal.

Vt = .1337 × Lg _____ = _____ ft.^{3}

Calculate Areas Covered by Tanks

A = (tank diameter/2)^2 × 3.14 = tank area

    1. Tank #1 = area   A1 = (_____ /2)^2 × 3.14 = _____ ft.^{2}

    2. Tank #2 = area   A2 = (_____ /2)^2 × 3.14 = _____ ft.^{2}

    3. Tank #3 = area   A3 = (_____ /2)^2 × 3.14 = _____ ft.^{2}

Total Area of Tanks: (At = area of tanks)

At = A1 + A2 + A3 = _____ ft.^{2}

Effective Containment Area: (Ae = effective containment area)

Ae = Ac − At = _____ ft.^{2}

Containment Wall Height Required (Hw = wall height)
(Vt = volume of largest tank)
(Ae = effective containment area)

Hw = Vt/Ae height of wall in feet

Hw = _____ ft.^{3}/ _____ ft.^{2} = _____ ft. height of wall

Dike Construction—Design Calculations for Single Tank Containment

Tank Data

1. Single tank, capacity: $Ct =$ _____ gal.

2. Tank diameter: $Dt =$ _____ ft.

3. Tank height: $Ht =$ _____ ft.

4. Available surface area within the containment wall (Ac = containment area):

   $Ac =$ _____ ft. × _____ ft. = _____ ft.^{2} inside walled area

Calculations

5. Tank area: A

   $A = (Dt/2)^2 \times 3.14$

   $A = ($ _____ $/2)^2 \times 3.14 =$ _____ ft.^{2}

6. Tank volume: Vt

   $Vt = At \times Ht$

   $Vt =$ _____ ft.$^2 \times$ _____ ft. = _____ ft.^{3}

7. Required containment volume: Vc

   $VC = 1.1 \times Vt$

   $VC = 1.1 \times$ _____ ft.$^3 =$ _____ ft.^{3}

8. Containment wall height, minimum required: Hw

   $Hw = Vc/Ac$

   $Hw =$ _____ ft.$^3 /$ _____ ft.$^2 =$ _____ ft.

9. Wall height _____ .

EXAMPLE 1

Dike Construction—Design Calculations for Single Tank Containment

Tank Data

1. Single tank, capacity: C_t = __2,000__ gal.
2. Tank diameter: D_t = __7.5__ ft.
3. Tank height: H_t = __7__ ft.
4. Available surface area within the containment wall (A_c = containment area):

     A_c = __10__ ft. × __10__ ft. = __100__ ft.^{2} inside walled area

Calculations

5. Tank area: A

     $A = (D_t/2)^2 \times 3.14$

     $A = ($ __7.5__ $/2)^2 \times 3.14 = $ __11.77__ ft.^{2}

6. Tank volume: V_t

     $V_t = A_t \times H_t$

     $V_t = $ __11.77__ ft.$^2 \times$ __7__ ft. = __82.4__ ft.^{3}

7. Required containment volume: V_c

     $V_c = 1.1 \times V_t$

     $V_c = 1.1 \times$ __82.4__ ft.$^3 = $ __90.64__ ft.^{3}

8. Containment wall height, minimum required: H_w

     $H_w = V_c/A_c$

     $H_w = $ __90.64__ ft.$^3/$ __100__ ft.$^2 = $ __.9__ ft.

9. Wall height __11 inches or 1 foot__ .

EXAMPLE 2

Dike Construction—Design Calculations for Single Tank Containment

Tank Data

1. Single tank, capacity: $Ct = \underline{\text{5,600}}$ gal.
2. Tank diameter: $Dt = \underline{\text{12}}$ ft.
3. Tank height: $Ht = \underline{\text{10}}$ ft.
4. Available surface area within the containment wall (Ac = containment area):

   $Ac = \underline{\text{15}}$ ft. $\times \underline{\text{15}}$ ft. $= \underline{\text{225}}$ ft.^{2} inside walled area

Calculations

5. Tank area: A

   $A = (Dt/2)^2 \times 3.14$

   $A = (\underline{\text{12}} /2)^2 \times 3.14 = \underline{\text{113}}$ ft.^{2}

6. Tank volume: Vt

   $Vt = At \times Ht$

   $Vt = \underline{\text{113}}$ ft.$^2 \times \underline{\text{10}}$ ft. $= \underline{\text{1,130}}$ ft.^{3}

7. Required containment volume: Vc

   $Vc = 1.1 \times Vt$

   $Vc = 1.1 \times \underline{\text{1,130}}$ ft.$^3 = \underline{\text{1,243}}$ ft.^{3}

8. Containment wall height, minimum required: Hw

   $Hw = Vc/Ac$

   $Hw = \underline{\text{1,243}}$ ft.$^3 / \underline{\text{225}}$ ft.$^2 = \underline{\text{5.5}}$ ft.

9. Wall height $\underline{\text{5.5 feet}}$.

EXAMPLE 3: Design Calculations for Multiple Tank Storage

Tank Data

1. Tank #1, capacity: C1 = __6,000__ gal. (largest volume)

   diameter: D1 = __12__ ft.

2. Tank #2, capacity: C2 = __2,000__ gal.

   diameter: D2 = __7__ ft.

3. Tank #3, capacity: C3 = __1,500__ gal.

   diameter: D3 = __7__ ft.

Sketch of Containment Area with Tanks Shown (Sketch Available Area to Place Tanks)

```
 _____
|                    | 20
|                    | -ft.
|_____|
     20 -ft.
```

Calculate Available Surface Area within the Containment Wall Area:
(AC = containment area)

AC = __20__ ft. × __20__ ft. = __400__ ft.^{2} inside wall area

Calculate Volume of Largest Tank Plus 10%: (Lg = gallons of largest tank)
(Vt = volume of largest tank)

Lg = C1 = __6,000__ gal. + __600__ (10%) gal. = __6,600__ gal.
Vt = .1337 × Lg __6,600__ = __882__ ft.^{3}

Calculate Areas Covered by Tanks

A = (tank diameter/2)^2 × 3.14 = tank area

1. Tank #1 = area A1 = (__12__ /2)^2 × 3.14 = __113__ ft.^{2}
2. Tank #2 = area A2 = (__7__ /2)^2 × 3.14 = __38.5__ ft.^{2}
3. Tank #3 = area A3 = (__7__ /2)^2 × 3.14 = __38.5__ ft.^{2}

Total Area of Tanks: (At = area of tanks)

At = A1 + A2 + A3 = __190__ ft.^{2}

Effective Containment Area: (Ae = effective containment area)

Ae = Ac − At = __210__ ft.^{2}

Containment Wall Height Required (Hw = wall height)
(Vt = volume of largest tank)
(Ae = effective containment area)

Hw = Vt/Ae height of wall in feet
Hw = __882__ ft.^{3}/ __210__ ft.^{2} = __4.2__ ft. height of wall

Back Flow
Device Data

Hose Connection Vacuum Breakers

Series 8

FOR BACKFLOW PREVENTION AS REQUIRED BY PLUMBING CODE CROSS-CONNECTION CONTROL—
Hose Supply Outlet that are not subject to continuous pressure

Watts 8A is furnished with an exclusive "non-hose thread faucets with portable hoses attached. Their purpose is to prevent the flow of contaminated water back into the safe drinking water supply . . . and installation requires no plumbing changes, device screws directly to the sill cock.

No. 8A is furnished with an exclusive "non-removable" feature that prevents removal from sill cock. This "Vandal Proof" feature is preferred by most Plumbing Codes.
(Pat. Nos. 3,459,443 and 3,171,423)

Applications

Series 8 can be used on a wide variety of installations such as service sinks, swimming pools, developing tanks, laundry tubs, wash racks, dairy barns, marinas and general outside gardening uses.

IMPORTANT: This valve should only be used in areas where spillage of water could not cause damage.

Available Models

No. 8A, 8, 8B, NF8, 8P,
No. S8C, 8AC, 8C, NF8-C, or 8BC
with chrome finished body.

Capacity

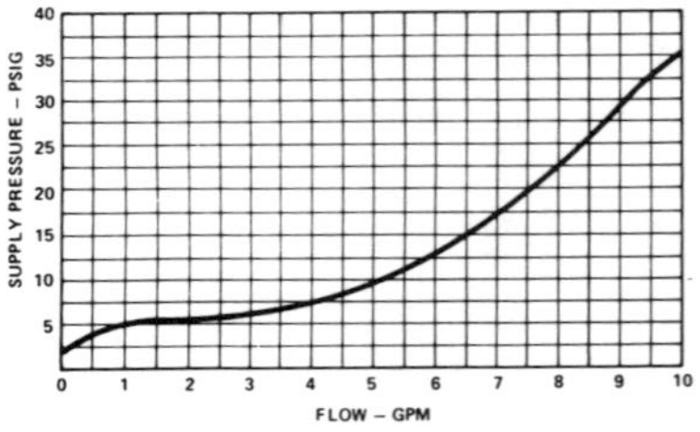

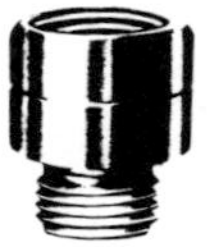

No. S8C ½" or ⅜"
Specially made for use with tub and shower hand spray sets. Installs without plumbing changes between shower-head and hose.
No. S8 plain brass.
Send for ES-S8.

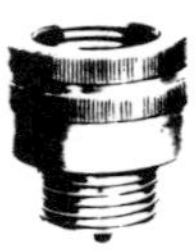

No. 8A
No. 8P

No. 8A for portable hoses attaches to hose thread faucets, requires no plumbing changes as it is screwed directly to the sill cock.
No. 8P same as 8A except body made of reinforced thermoplastic.
Send for ES-8 and ES-8P.

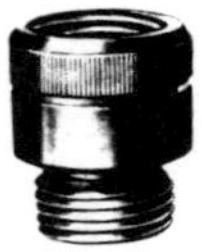

No. 8
Similar to No. 8A Series except it is furnished without the "Non-removable" or draining feature.
Send for ES-8.

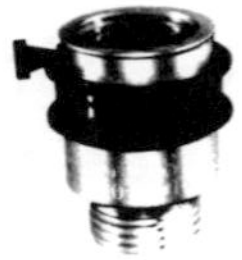

No. NF8
No. NF8 for wall and yard hydrants. Permits manual draining for freezing conditions.
Send for ES-8

Operation of Watts No. 8A Series

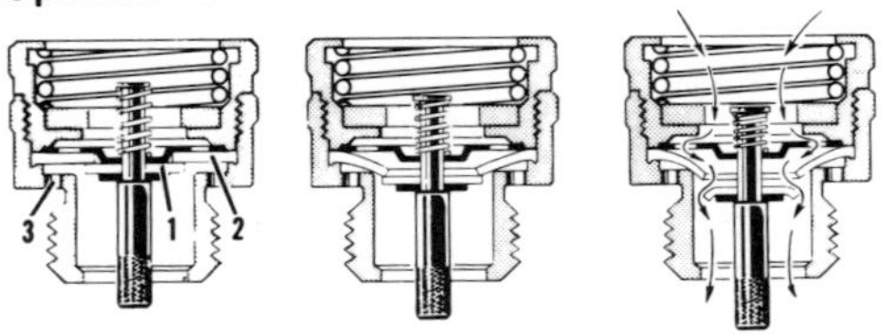

No. 8A in closed position with supply valve shut off disc (1) seats against diaphragm (2). Atmospheric ports are open (3) during no flow.

Before flow begins, atmospheric ports are sealed off before lower disc opens to permit flow.

Lower disc opens away from atmospheric diaphragm seal allowing flow through the valve with slight pressure drop.

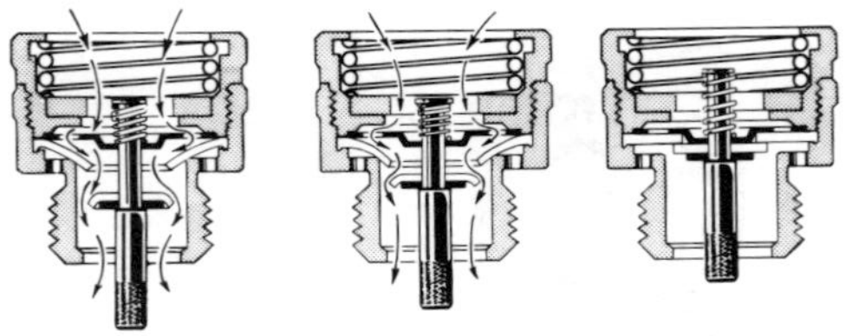

Fully opened valve, illustrating poppet action to provide high capacity with minimum pressure drop through valve.

With reduction in water supply pressure, disc returns closer to diaphragm. Diaphragm seals off atmospheric ports preventing unnecessary leakage.

With loss of water supply, disc seals tightly against diaphragm preventing back-siphonage or backflow of water and opens atmospheric ports.

Construction

No. 8A - Furnished with exclusive "Non-removable" feature and standardly equipped to allow sill cock to be drained.
Note: Device should only be installed on approved sill cocks containing at least four full threads and is non-removable once installed.

No. 8P - Furnished with exclusive patented "Non-removable" feature and standardly equipped to allow sill cock to be drained. Constructed of durable, corrosion- resistant, reinforced thermoplastic. Tamper-proof feature. Patent No. 4821763.

No. 8 - Similar to the No. 8A except it is furnished without the "Non-removable" or draining feature.

No. 8B - Furnished with breakaway set screw to provide a tamper-resistant installation and standardly equipped to allow sill cock to be drained.

No. NF8 - Especially made for wall and yard hydrants. Permits manual draining for freezing conditions.

DRAINAGE FEATURES TO PREVENT FREEZING
Watts No. 8A, 8B and 8P are standardly equipped to allow sill cocks to be drained. To do this, remove hose coupling and lightly pull knurled tip of stem at outlet of valve to allow drainage of collected water.

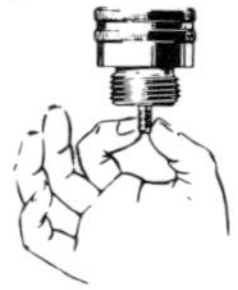

Note: Do not use No. 8, 8A, 8B, 8P Hose Bibb Vacuum Breakers on frost-free hydrants. Specify No. NF8.

Installation

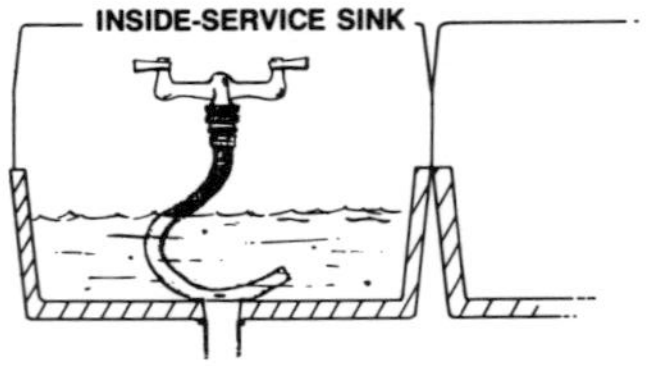

WHAT IS BACK-SIPHONAGE?
A reversal of normal flow in the system caused by a negative pressure (vacuum or partial vacuum) in the supply piping.

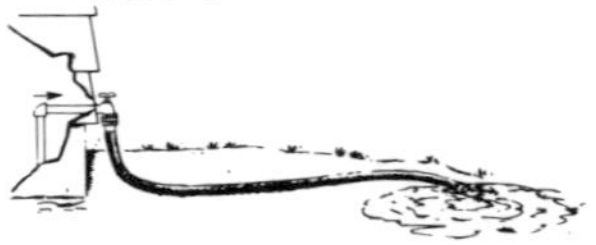

Dimensions-Weight (are approximate)

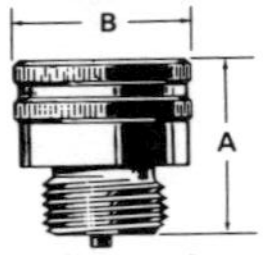

No.	Size	A	B	Weight
8,8C, 8B, 8BC	¾" HT	1½"	1⅜"	¼ lb.
8A, 8AC	¾" HT	1½"	1½"	¼ lb.
NF8, NF8C	¾" HT	2"	1½"	⅓ lb.
8P	¾" HT	1 11/16"	1⅜"	1½ oz.

Reduced Pressure Zone Backflow Preventers

909 OSY RW

Fused epoxy coated inside and out

Series 909
Sizes 3/4" – 2"

"Engineered for high capacity relief"

Patent # 4,241,752

909QT-S

To prevent back-siphonage and backpressure of con-taminated water into the safe drinking water supply, when installed at each high hazard cross-connection.

Use Series 909 for backflow protection in cross-connec-tion control and containment at the service entrance. The 909 high capacity relief incorporates the "air-in/water-out" principle and substantially improves the relief valves dis-charge performance. The emergency condition of com-bined back-siphonage and backpressure with both checks fouled can reduce the effectiveness of a standard RPZ backflow preventer. Standardly furnished with NPT body connections and quarter-turn, full port, resilient seated, bronze ball valve shut-offs No. 909QT. Sizes 3/4" and 1" have Tee handle shut-offs.

Available Models

Suffix
QT - with quarter-turn ball valve, full port, resilient seated ball valve shut-offs
S - with bronze strainer
HW - with stainless steel check modules for hot water and harsh water conditions
LF - without shut-off valves

Prefix
U - with integral body unions (3/4" and 1" only)
FAE - with flanged adapter ends (1 1/4", 1 1/2", 2" only)

Features
- Quarter-turn ball valve shut offs (Standard)
- Replaceable seats and seat discs
- Modular design
- Simple and economical service servicing
- No special tools required for servicing
- High capacity relief protection against combined back-siphonage/backpressure backflow

Standards (see page 3)

Pressure-Temperature
Series 909 suitable for supply pressure up to 175 psi and water temperatures up to 140°F continuous and 180°F in-termittent. Suffix HW stainless steel check modules suit-able for supply pressure up to 175 psi and water tempera-ture up to 210°F for harsh water conditions specify PC model

Connections
3/4" - 1" 909QT has NPT female threaded body connections. 1 1/4" - 2" 909QTM1 has NPT male threaded body connections.

Dimensions-Weights (approximate)

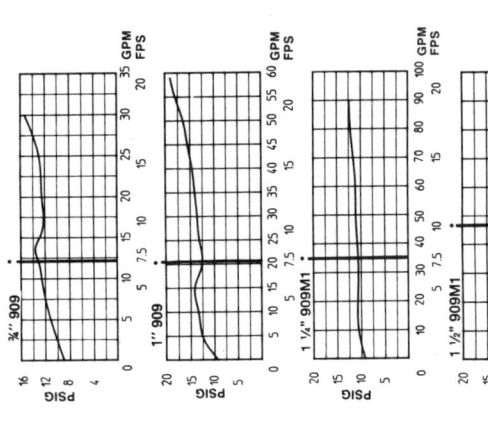

909QT-S

Size	A	B	C	D	E	F	G	H	Weight (lbs.) Less Strainer	With Strainer
3/4"	21 1/4	17 1/4	7 5/16	4	4 3/4	11 1/4	6 3/4	3 7/8	15	15 5/8
1"	22 1/4	17 3/16	7 5/16	4	4 3/4	13	7 1/2	3 7/8	15	17 1/2
1 1/4"M1	25 5/8	20 5/8	11 3/4	5	6 5/8	14	7 1/2	5 1/4	40	47 3/4
1 1/2"M1	27 3/16	20 5/8	11 3/4	5	6 5/8	14	7 1/2	5 1/4	44	47 3/4
2"M1	29 3/4	23 5/8	11 3/4	5	6 5/8	14	7 1/2	5 1/4	40	47 3/8

For more information, send for ES-909S.

Series 909
Sizes 2 1/2" – 10"

Series 909 2 1/2" - 10" provide backflow protection in cross-connection control and containment with its unique patented design incorporating the "air-in/water-out" principle.

Available Models

Suffix
NRS RW - with non-rising stem resilient seated gate valves
S-FDA - with FDA approved epoxy coated strainer
BB - with bronze body (2 1/2", 3")
OS&Y RW - with UL/FM resilient seated outside stem and yoke gate valves
QT - with quarter-turn, full port, resilient seated ball valve shut-offs
QT-FDA - with FDA approved epoxy coated ball valve shut-offs
LF - without shut-offs.
Note: The installation of a drain line is recommended. When installing a drain line, an air gap is necessary (see page 5).

Capacity
As compiled from documented Foundation for Cross-Connection Control and Hydraulic Research at the University of Southern California lab tests.
*Typical maximum mechanical/irrigation system flow rate (7.5 feet per second)

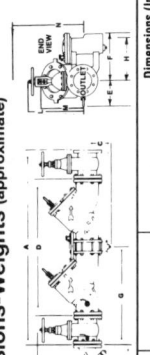

Features
- Replaceable bronze seats
- Resilient seated gate valve shut-offs
- No special tools required for servicing
- FDA approved epoxy coated check and relief valves (inside and out)

Standards (see page 3)

Pressure-Temperature
Suitable for supply pressure up to 175 psi and water tem-peratures up to 110°F continuous and 140°F intermittent.

Materials
Series 909 sizes 2 1/2" - 10" have FDA approved epoxy coated cast iron check valve bodies with bronze check seats and stainless steel relief valve seat.

Note: The installation of a drain line recommended. When installing a drain line, and air gap in necessary (see page 5).

BACKFLOW PREVENTION FOR HIGH HAZARD CROSS-CONNECTION and CONTAINMENT

INSTALLATIONS WITH CONTINUOUS PRESSURE

Capacity
As compiled from documented Foundation for Cross-Connection Control and Hydraulic Research at the University of Southern California lab tests.
*Typical maximum mechanical/irrigation system flow rate (7.5 feet per second)

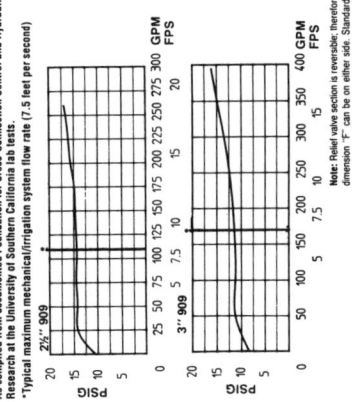

Dimensions-Weights (approximate)

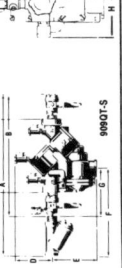

END VIEW

Note: Relief valve section is reversible; therefore, dimension "F" can be on either side. Standardly furnished as shown. **Specify RH** if relief valve is to be supplied on the opposite side from standard.

Size	A	B NRS	B OS&Y	B QT	C	D	E	F	G	H	I	Dimensions (Inches) *J	K	*L	*M	*N	NRS	OS&Y*	QT
2 1/2"	41 1/8	11 3/8	15 7/8	7	5 1/4	26 5/8	4	9 1/8	20 5/8	7 5/8	10	10	6 1/2	14	11	20	195	198	182
3"	42 1/8	12 3/8	18 3/8	7	5 1/4	26 5/8	5	9 1/8	21 3/4	7 7/8	10 1/4	10	7	14	11	20	225	230	190
4"	55	15 3/8	23 3/4	10	6	37	6	14 3/8	27 7/8	12 1/2	12	12	8 1/2	17	14	25	455	470	352
6"	65 1/8	19 3/4	32 1/2	15	6	44 1/2	11	14 3/8	32 1/4	15 1/8	18 1/2	20	13 1/2	21	16	34	718	798	762
8"M1	78 1/2	24 1/2	39 1/4	19	9 1/4	55 1/4	11 1/4	19 1/4	39 1/4	17 7/8	21 1/2	22 3/4	18 1/2	26	21	41	1,350	1,456	2286
10"M1	93 5/8	29 3/4	48	22	9 3/4	67 3/4	12 1/2	21	46 5/8	19 1/8	26	28	18 1/2	32	21	52	2,160	2,230	3116

Weight (lbs.)

*J, L, M, N dimension are clearance required for servicing (in. inches).
For more information, send for ES-909L.
Inquire with governing authorities for local requirements

Atmospheric Vacuum Breakers
Series 288A ¼"- 3"

BACKFLOW PREVENTION FOR HIGH HAZARD CROSS-CONNECTIONS
— As required by plumbing code cross-connection control — Installations that are not subject to continuous pressure

Designed to prevent back-siphonage of contaminated water into a safe drinking water supply.

Watts 288A features a lightweight, durable "disc float" which closes the atmosphere vent to **prevent spilling under all rates of flow.** Suitable for temperatures up to 180°F. Therefore, they are ideally recommended for low flow installations such as laboratory equipment which use a small amount of water. They also contain a durable silicone disc which has high heat and water hammer shock resistance and assures tight-seating with the lightest of seating contact.

Applications

Vacuum breakers are used in the water supply line to all types of equipment and fixtures connected to a safe drinking water supply system and on which the water supply enters the fixture or device below its flood rim. Typical of this type of equipment are clothes and dishwashers, beverage dispensers, chemical and process tanks and similar devices.

Standards (See page 3)

Materials

Sizes 1/4" through 1" are available in either plain brass or polished chrome finish. Durable silicone disc, high heat resistance, Sizes 1/4" through 3" standardly furnish with plain brass finish.

Pressure-Temperature

Maximum temperature 180°F at 125 lbs. working pressure Note: This device shall not be subjected to continuous pressure for more than 12 hours, per A.S.S.E. Std. 1001

Dimensions-Weights (approximate)

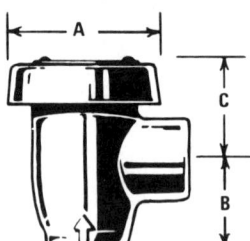

Capacity

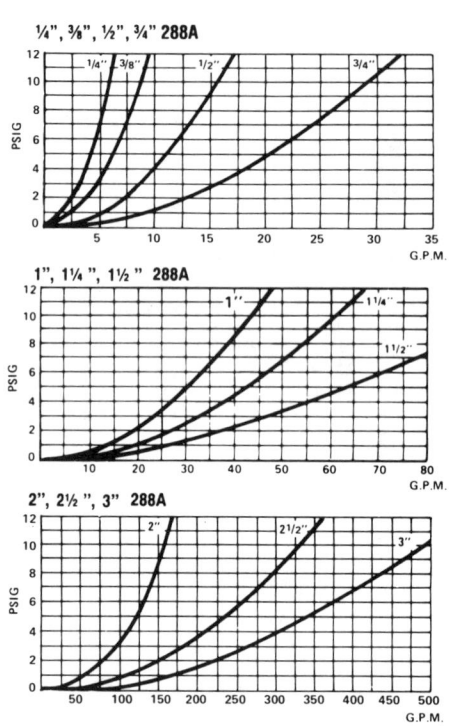

For Additional infomation send for ES-288A
For information on Irrigation vacuum breakers send for IS/ES-188A

No.		Size	Dimensions			Weight
Plain Brass	Polished Chrome		A	B	C	
288A	288A-C	¼"	1¹¹/₁₆"	1¹/₁₆"	1³/₁₆"	³/₈ lb.
288A	288A-C	³/₈"	1¹¹/₁₆"	1¹/₁₆"	1³/₁₆"	³/₈ lb.
288A	288A-C	½"	2"	1³/₁₆"	1⁷/₁₆"	½ lb.
188/288A	288A-C	¾"	2¼"	1½"	1⁷/₈"	1⅛ lbs.
188/288A	288A-C	1"	2⁷/₈"	1¹¹/₁₆"	2⅛"	1¾ lbs.
188/288A		1¼"	2⁷/₈"	1¹³/₁₆"	2⅛"	2⅛ lbs.
188/288A		1½"	3⅝"	2³/₁₆"	2⁷/₁₆"	3⅝ lbs.
188/288A		2"	4⅛"	2½"	2⁷/₈"	5¼ lbs.
288A		2½"	6³/₈"	3"	4³/₈"	16 lbs.
288A		3"	6³/₈"	3³/₈"	4⅝"	17⅛ lbs.

Concrete Pad Dimensions

Dimension	Pad Size (ft×ft)						
	20×30	30×40	40×53	50×63	60×75	70×85	80×100
A	20'-0"	30'-0"	40'-0"	50'-0"	60'-0"	70'-0"	80'-0"
B	10'-0"	10'-0"	13'-0"	13'-0"	15'-0"	15'-0"	20'-0"
C	30'-0"	40'-0"	53'-0"	63'-0"	75'-0"	85'-0"	100'-0"
D	20'-0"	30'-0"	40'-0"	50'-0"	60'-0"	70'-0"	80'-0"
E	10'-0"	15'-0"	20'-0"	25'-0"	30'-0"	35'-0"	40'-0"
F	14"	17"	18"	18"	18"	21"	24"
G	14"	17"	18"	21"	24"	27"	30"
H	16.5"	21"	23"	27"	31"	35"	38"
*I	0"	0"	0"	0"	0"	0"	0"
J	8"	11"	12"	15"	18"	21"	24"
K	5.5"	7"	7"	9"	11"	13"	16"
L	8"	11"	12"	15"	18"	21"	24"
M	12"	18"	18"	18"	18"	18"	18"
N	18"	30"	30"	30"	30"	30"	30"
O	18"	30"	30"	30"	30"	30"	30"
P	4.5"	5.0"	5.0"	5.5"	5.5"	5.5"	6.0"
Q	3.5%	3.0%	2.5%	2.5%	2.5%	2.5%	2.5%
R	⅜"	⅜"	⅜"	½"	½"	½"	½"
S	10'	10'	13'	13'	15'	14'	13'
T	7'	10'	13'	13'	15'	14'	13'
**U	750 Gal.	1,500 Gal.	2,900 Gal.	4,750 Gal.	7,700 Gal.	10,600 Gal.	19,000 Gal.
V	21½"	33½"	33½"	33½"	33½"	33½"	33½"
W	19½"	31½"	31½"	31½"	31½"	31½"	31½"
X_1	24"	36"	36"	36"	36"	36"	36"
X_2	19⅞"	31⅞"	31⅞"	31⅞"	31⅞"	31⅞"	31⅞"
Y_1	26"	38"	38"	38"	38"	38"	38"
Y_2	21⅞"	33⅞"	33⅞"	33⅞"	33⅞"	33⅞"	33⅞"

NOTES:
* "I" can be increased to provide increased containment volume; If "I" is changed, add the new "I" value to "G," "H," "J," and "K" dimensions.
** "U" = Total Containment Section Volume, Gallons, displacement volume of tanks, and equipment must be subtracted to determine net usable volume to meet EPA requirments.

Concrete Mix Data

Watertight Concrete Mix Designs—Approximate Values						
(0.40–0.45 Water-Cement Ratio)						
Maximum Coarse Aggregate Size (Inches)	Total (1) Mixing Water Lb. (Gals.) Per Cu. Yd. 0.40/0.45 W-C Ratio	Entrained Air Content % by Volume Concrete	Cement (2) Lb. (Bags) Per Cu. Yd. 2.60 Fineness Modulus	Fine Aggregate Lb./Cu. Yd. 2.60 Fineness Modulus	Coarse Aggregate Lb./Cu. Yd.	Total Weight Lb./Cu. Yd. .040/.045
½	195 (24)/225 (27)	7.0	650 (6.9)	1320	1540	3705/3735
¾	175 (21)/205 (25)	6.0	615 (6.5)	1245	1730	3765/3795
1	160 (19)/190 (23)	6.0	575 (6.2)	1185	1860	3780/3810
1½	145 (19)/175 (21)	5.5	540 (5.7)	1165	1970	3820/3850

Note: (1) Available surface water in coarse aggregate estimated @ 1.5% M.C. Available surface water in fine aggregate eliminated @ 3.5% M.C. Gravel w/some crushed aggregate assumed to decrease water required by 20 lb/cu. yd. (2) Type IIA or IA, or Type II or I portland cement with air-entrainment additive.

Concrete and Steel Components Estimates							
Pad Size		Concrete (1)		Reinforcing Steel (2)			
W × L	Total Volume	Pad Thickness		⅜" Dia.	½" Dia.	Steel Pattern Spacing (3)	
Ft. × Ft.	Cu. Yd.	Containment (Inches)	Wash (Inches)	Ft. (Lb.)	Ft. (Lb)	Pads & Walls (Inches)	Sump (Inches)
20 × 30	12	6.0	4.5	2,000 (750)		12 × 12	6 × 6
30 × 40	24	6.0	5.0	3,800 (1,430)		12 × 12	6 × 6
40 × 53	42	6.0	5.0	6,250 (2,350)		12 × 12	6 × 6
50 × 63	66	6.0	5.5		8,850 (5,930)	12 × 12	6 × 6
60 × 75	97	6.0	5.5		12,220 (8,190)	12 × 12	6 × 6
70 × 85	128	6.0	5.5		15,675 (10,500)	12 × 12	6 × 6
80 × 100	177	6.0	6.0		20,480 (13,720)	12 × 12	6 × 6

Note: (1) Concrete volumes computed at 10% overrun for uneven soil; (2) Steel includes 5% joint overlap allowance; overlap all steel joints—20 diameters for 12" minimum; (3) Sump area steel at 6" × 6" spacing in all sidewalls and base with an 18"-extension into the concrete pad.

Concrete Cost Data

		1988 Chemical Handling Concrete Pad Cost Estimates						
Item	**Unit Cost**	**Concrete Pad Size (ft. × ft.)**						
Concrete—Type II Volume (Cu. Yd.) Cost Range	Unit Costs (Low/High)	20 × 30 12 (Low/High)	30 × 40 24 (Low/High)	40 × 53 42 (Low/High)	50 × 63 66 (Low/High)	60 × 75 97 (Low/High)	70 × 85 128 (Low/High)	80 × 100 177 (Low/High)
Concrete Cost 6.0 Bag Mix 7.0 Bag Mix	($/cu. yd.) $45/55 $50/60	$540/660 $600/720	1,080/1,320 1,200/1,440	1,890/2,310 2,100/2,520	2,970/3,630 3,300/3,960	4,365/5,335 4,850/5,820	5,760/7,040 6,400/7,680	7,965/9,735 8,850/10,620
Air Entrainment 6.0 Bag Mix 7.0 Bag Mix	($/cu. yd.) $0.75/1.25 $1.00/1.50	$9/15 $12/18	18/30 24/36	32/53 42/63	50/82 66/99	74/121 97/145	96/160 128/192	133/221 177/266
Plastic Admixture 6.0 Bag Mix 7.0 Bag Mix	($/cu. yd.) $3.75/4.50 $4.50/5.00	$45/54 $54/60	90/108 108/120	158/185 185/210	248/297 297/330	365/437 437/485	480/576 576/640	664/797 797/885
Delivery Cost Beyond 3–5 Miles Base Zone Cost per Mile	($/cu. yd.) $.20/.30	$2.40/3.60	4.80/7.20	8.40/12.60	13.20/19.80	19.40/29.10	25.60/38.40	35.40/53.10
Reinforcing Steel (Weight) (Lb.) Cost ($/Lb.)	$.20/.30	(750) $150/225	(1,430) 286/429	(2,350) 473/710	(5,930) 1,186/1,780	(8,190) 1,638/2,457	(10,500) 2,100/3,150	13,720) 2,744/4,116
Forming Labor ($/sq. ft.) ($/cu. yd.)	$.20/.30 $12.00/18.00	$144/216	288/432	504/756	792/1,188	1,165/1,745	1,536/2,304	2,124/3,186
Finishing Labor ($/sq. ft.) ($/cu. yd.)	$.20/.30 $12.00/18.00	$144/216	288/432	504/756	792/1,188	1,165/1,745	1,536/2,304	2,124/3,186
Total Cost @ Base Delivery Zone 6.0 Bag Mix 7.0 Bag Mix		$1,034/1,390 $1,106/1,459	2,055/2,758 2,199/2,896	3,569/4,783 3,816/5,028	6,051/8,185 6,446/8,565	8,791/11,869 9,371/12,426	11,534/15,572 12,302/16,308	15,789/21,294 16,851/22,312

Tank Size vs. Base Area and Volume

		Tank Size vs. Base Area and Volume										
		Tank Depth (Ft.)										
Tank Dia. (Ft.)	Base Area (Sq. Ft.)	1	2	3	4	5	10	12	14	16	18	20
		Tank Volume (Gal.)										
15	176.7	1,325	2,650	3,975	5,300	6,625	13,250	15,900	18,550	21,200	23,850	26,500
14	153.9	1,154	2,308	3,462	4,616	5,770	11,540	13,848	16,156	18,464	20,772	23,080
13	132.7	995	1,990	2,985	3,980	4,975	9,950	11,940	13,930	15,920	17,910	19,900
12	113.1	848	1,696	2,544	3,392	4,240	8,480	10,176	11,872	13,568	15,624	16,960
11	95.0	713	1,426	2,139	2,852	3,565	7,130	8,556	9,982	11,408	12,834	14,260
10	78.5	589	1,178	1,767	2,356	2,945	5,890	7,068	8,246	9,424	10,602	11,780
9	63.6	477	954	1,431	1,908	2,385	4,770	5,724	6,678	7,632	8,586	9,540
8	50.3	377	754	1,131	1,508	1883	3,770	4,524	5,278	6,032	6,786	...
7	38.5	289	578	867	1,156	1,445	2,890	3,468	4,046	4,624	...	...
6	28.3	212	424	636	848	1,060	2,120	2,544	2,968	...	...	...
5	19.6	147	294	441	588	735	1,470	1,764	...	...	...	...

Containment Pad Size vs. Tank Size and Number

Tank Size	CVD	Number of Tanks							
		1	2	3	4	5	6	7	8
15×20	2	36×60	39×60	42×60	45×60	48×60	51×60	54×60	57×60
		43×50	47×50	50×50	54×50	57×50	61×50	65×50	68×50
	3	24×60	27×60	30×60	33×60	36×60	39×60	42×60	45×60
		29×50	32×50	36×50	39×50	43×50	46×50	50×50	53×50
	4	25×40	20×60	23×60	26×60	29×60	32×60	35×60	38×60
		20×50	24×50	27×50	31×50	34×50	38×50	42×50	45×50
14×20	2	32×60	34×60	37×60	39×60	42×60	44×60	47×60	50×60
		38×50	41×50	44×50	47×50	50×50	53×50	56×50	60×50
	3	21×60	23×60	26×60	29×60	31×60	34×60	36×60	39×60
		25×50	28×50	31×50	34×50	37×50	40×50	43×50	47×50
	4	23×40	18×60	21×60	23×60	26×60	28×60	31×60	34×60
		19×50	22×50	25×50	28×50	31×50	34×50	37×50	40×60
13×20	2	27×60	30×60	32×60	34×60	36×60	38×60	41×60	43×60
		33×50	35×50	38×50	41×50	43×50	46×50	49×50	51×50
	3	18×60	20×60	22×60	25×60	27×60	29×60	31×60	34×60
		22×50	24×50	27×50	30×50	32×50	35×50	38×50	40×50
	4	20×40	16×60	18×60	20×60	22×60	25×60	27×60	29×60
		16×50	19×50	21×50	24×50	27×50	30×50	32×50	35×50
12×20	2	23×60	25×60	27×60	29×60	31×60	33×60	35×60	37×60
		28×50	30×50	32×50	35×50	37×50	39×50	42×50	44×50
	3	16×60	18×60	19×60	21×60	23×60	25×60	27×60	29×60
		19×50	21×50	23×50	25×50	28×50	30×50	32×50	34×50
	4	17×40	20×40	23×40	17×60	19×60	21×60	23×60	25×60
		23×30	16×50	18×50	21×50	23×50	25×50	27×50	30×50
11×20	2	20×60	22×60	23×60	26×60	26×60	28×60	30×60	31×60
		24×50	26×50	28×50	30×50	32×50	34×50	36×50	38×50
	3	16×50	18×50	20×50	21×50	23×50	25×50	27×50	29×50
		20×40	22×40	24×40	27×40	29×40	31×40	34×40	36×40
	4	15×40	17×40	19×40	22×40	24×40	26×40	29×40	31×40
		19×30	23×30	26×30	29×30	32×30	35×30	38×30	42×30
10×20	2	16×60	18×60	19×60	20×60	22×60	23×60	24×60	25×60
		20×50	21×50	23×50	24×50	26×50	27×50	29×50	30×50
	3	13×50	15×50	16×50	18×50	19×50	21×50	22×50	24×50
		16×40	18×40	20×40	22×40	24×40	26×40	28×40	30×40
	4	19×25	14×40	16×40	18×40	20×40	22×40	24×40	26×40
		16×30	19×30	21×30	24×30	26×30	29×30	32×30	34×30
9×20	2	16×50	17×50	18×50	20×50	21×50	22×50	24×50	25×50
		20×40	21×40	23×40	24×40	26×40	28×40	29×40	31×40
	3	13×40	15×40	16×40	18×40	20×40	21×40	23×40	24×40
		17×30	20×30	22×30	24×30	26×30	28×30	30×30	32×30
	4	16×25	18×25	21×25	15×40	16×40	18×40	19×40	21×40
		13×30	15×30	17×30	19×30	21×30	24×30	26×30	28×30
8×18	2	14×40	15×40	17×40	18×40	19×40	20×40	22×40	23×40
		19×30	· 20×30	22×30	24×30	25×30	27×30	29×30	30×30
	3	15×25	17×25	16×30	13×40	15×40	16×40	17×40	18×40
		13×30	14×30	16×30	17×30	19×30	21×30	22×30	24×30
	4	14×20	16×20	19×20	14×30	16×30	18×30	19×30	21×30
		18×15	13×25	15×25	17×25	19×25	21×30	23×25	25×25

Suppliers of Approved Equipment and Services

Tanks

Crown Rotational Molded Products
Box 577
Marked Tree, AR 72365
501-358-3400

Precision Tank & Equipment
P.O. Box D
Virginia, IL 62691
217-452-7228

Certified Equipment
2500 Richards Lane
Springfield, IL 62705
217-525-1433

Murray Equipment
2515 Charleston Place
Ft. Wayne, IN 46808
800-348-4753

Snyder Ind. Inc.
P.O. Box 4583
Lincoln, NE 68504
402-467-3247

Chemical Tank Distributing
28600 Quinn Road
North Liberty, IN 46554
219-656-8905

Raven Industries
P.O. Box 1007
Sioux Falls, SD 57117

Farm Chem Corp.
Box 300
Floyd, IA 50435
800-247-1854

Solar Plastics
732 30th Avenue, SE
Minneapolis, MN
612-331-8636

Snyder Industries
4620 Fremont Street
Lincoln, NE 68504
402-467-5221

Assmann Corp.
300 N. Taylor Road
Garrett, IN 46738
219-357-3181

Modern Welding Company
P.O. Box 2265
72 Waldo Street
Newark, OH 43055
614-344-9425

Meters and Pumps

Liquid Controls Corp.
P.O. Box 101
No. Chicago, IL 60064
312-689-2400

Farm Chem Corp.
Box 300
Floyd, IA 50435
800-247-1854

Chemical Tank Distributing
28600 Quinn Road
North Liberty, IN 46554
219-656-8905

Murray Equipment
2515 Charleston Place
Ft. Wayne, IN 46808
800-348-4753

Great Plains Industries
1711 Longfellow Lane
Wichita, KS 67207
316-686-7361

Tokheim Corp.
1606 Wabash Avenue
Ft. Wayne, IN 46801
219-423-2532

MP Pumps
515 Lycaste
Detroit, MI 48214
313-822-0240

Nozzles and Ball Valves

Murray Equipment
2515 Charleston Place
Ft. Wayne, IN 46808
800-348-4753

Farm Chem Corp.
Box 300
Floyd, IA 50435
800-247-1854

Chemical Tank Distributing
28600 Quinn Road
North Liberty, IN 46554
219-656-8905

Sealants, Coatings, and Caulking

Gibson-Homans Company
1755 Enterprise Parkway
Twinsburg, OH 44087
216-425-3255

Lauren Mfg. Co.
2228 Reiser Avenue SE
New Philadelphia, OH 44663
216-339-3373

The Euclid Chemical Co.
19218 Redwood Road
Cleveland, OH 44110

Wisconsin Protective
   Coatings
614 Elizabeth St.
Green Bay, WI 54302
414-437-6561

Steelcote Manufacturing
1 Steelcote Square
St. Louis, MO 63103
314-771-8053

Prefab Dikes and Buildings

CHEM-Collector Pad
9601 N. Allen Rd.
Peoria, IL 61615
800-747-5211

Hunter Agri Sales, Inc.
Box 2
Coatsville, IN 46121
317-386-7277

Safety Storage, Inc.
2380 S. Bascom Avenue
Campbell, CA 95008
408-559-3901

Mod U Tank Inc.
29-24 40th Ave.
Long Island City, NY 11101
718-392-1112

Waste Disposal

Ross Environmental
394 Giles Road
Grafton, OH 44044
216-748-2171

Safety Supplies and Accessories

Spectrum Technologies, Inc.
12010 S. Aero Drive
Plainfield, IL 60554
815-436-4440

J. V. Mfg. Co.
963 Ashwaubenow Street
Green Bay, WI 54304
800-334-9092

Safety Equipment & Supply Co.
5400 Distribution Drive
P.O. Box 5113
Ft. Wayne, IN 46895
219-482-8441

References

Bucklin, R. A., D. Bottcher, and S. Dwinell, "Evaporation/Degradation System for Pesticide Equipment Rinse Water," Bulletin 242, Florida Cooperative Extension Service. Gainesville: University of Florida, 1987. 22 p.

CIBA-GEIGY Guide to Handling Bulk Pesticides, 1990.

Effects of Substances on Concrete and Guide to Protective Surfaces, 1986. PCA Bulletin IS001.06T. Skokie, IL: Portland Cement Association. 21 p.

Gilding, T. J., editor. "Managing Pesticide Wastes: Recommendations for Action," Summary of National Conferences and Workshops on Pesticide Waste Disposal. Washington, D.C.: National Agricultural Chemical Association, July 1988. 85 p.

Journal of Protective Coatings. Buyers Guide.

Noyes, R. T. and D. W. Kammel, "Modular Concrete Wash/Containment Pad for Agricultural Chemicals," International ASAE Meeting, New Orleans, LA. December 12–15, 1989.

Professional Lawn Care Association of America. *Technical Resource Manual*, 1986.

Rester, Darryl, "Waste Water Recycling," Paper No. AA86-001, 1986 Joint Technical Session of National Agricultural Aviation Association and American Society of Agricultural Engineers, Acapulco, Mexico, December, 1986. 11 p.

Steelcote Manufacturing Co./Engineering Service.

"Thickness Design for Concrete Highway and Street Pavements," PCA Engineering Bulletin EB109.01P, Skokie, IL: Portland Cement Association, 1984. 47 p.

"Watertight Concrete," PCA Bulletin IS002.03T, Skokie, IL: Portland Cement Association, 1975. 4 p.

Wiss, Janney, Elstner Assoc., Inc., "Investigations of the Use of Lauren Sealers on Portland Cement Concrete." April 14, 1988.

Index